MAKE MORE MONEY:
12 PROFIT PILLARS FOR HVAC CONTRACTOR SUCCESS

MAKE MORE MONEY:
12 PROFIT PILLARS FOR HVAC CONTRACTOR SUCCESS

BY

Scott Ritchey

with

Gary Kerns

www.bookstandpublishing.com

Published by
Bookstand Publishing
Morgan Hill, CA 95037
4648_6

Copyright © 2018 by Scott Ritchey and Gary Kerns
All rights reserved. No part of this publication may be reproduced or transmitted in any form or by any means, electronic or mechanical, including photocopy, recording, or any information storage and retrieval system, without permission in writing from the copyright owner.

ISBN 978-1-63498-727-1

Library of Congress Control Number: 2018949180

Printed in the United States of America

Dedication

This book is dedicated to every HVAC contractor who wants to run a successful business. The business and financial information outlined in this book are guaranteed to work for any size contractor.

Table of Contents

Dedication v

Testimonials ix

Foreword xi

Introduction xiii

Chapter 1: It's a Jungle Out There 1

Chapter 2: Profits through Market Cap Pricing 9

Chapter 3: Cost Control: Managing Your Overhead 21

Chapter 4: Markups, Margin, Pricing, and Market Caps ... 37

Chapter 5: Your Replacement Business Opportunity 53

Chapter 6: Managing Multiple Businesses 65

Chapter 7: We Don't Know What We Don't Know: Financials 75

Chapter 8: Ratios: What Your Financials Tell You 95

Chapter 9: The Service Tech Goldmine 117

Chapter 10: Technician Training and Compensation Ideas 137

Chapter 11: Knowledge is Confidence: In Their Own Words 153

Chapter 12: Time to Get Started and Just Do It! 165

About the Authors 174

Testimonials

"I've worked with Scott for more than 25 years, and his books and strategies got me to where I am today. I came to him to learn how to compete with the big boys, and now I am one of them."

— Herb Hovey, Owner, HH Hovey

"If I had a friend who was struggling with their business like I was, I would say the same thing that was said to me: "I know a guy who will change your life. And that guy is Scott Ritchey."

— Rich Mullins, Owner, H2O Plumbing

"With the aid of these strategies, I took my business from $600,000 to its current multimillion-dollar volume, and I'm well on my way to my goal of changing not only my family tree but the family tree of my employees as well, as I am able to offer them more than most companies because of my higher profits. Scott Ritchey gave me the business know-how I needed to grow my business."

—Brian Schneider, Owner, Allegiance Heating & Air

"After applying Scott's strategies, my growth continued. In 2015, I did $1.2 million in sales. Last year, I did $1.5 million. This year, $1.7 million. That is a nice steady rate of growth, and we are currently preparing to expand into a 3,200-square-foot space."

—Werner Vankleef, Owner, Vankleef's Heating and Air

"We're showing growth every year. We're up to an average of $200,000-plus per service truck, which is better than the industry benchmark. We started as a 100% residential company, but today we have a very strong commercial business as well; that's where the big money is. That's what I learned, to go where the biggest profit was. I would say to other business owners, you need someone like Scott Ritchey."

—Van Edwards, Owner, Edwards Refrigeration

x

Foreword

Throughout my many years of talking with small business owners, I have often shared a simple, one-line statement that I believe speaks to the heart of what they face: Small business is tough. In my 25+ years leading the sales team at an employee-owned HVAC distribution company, I worked with contractors throughout Illinois and Iowa.

During this time, I witnessed many of the issues that business owners face while simply striving to earn a profit that exceeds their risks—let alone trying to grow their companies to provide opportunities for their families and their employees. In the past two years I have gained firsthand knowledge of these issues as I have taken an ownership position at a residential service/replacement company.

It's no secret that owning a business comes with risk. Many of the companies with which I have worked have struggled immensely—some of them for many years. Sadly, I have also watched some of these companies fail. And I've seen the impact that has on their owners.

At the same time, I have worked with companies that have thrived—providing their owners with incomes and lifestyles that far exceed anything they could have hoped for as an employee. What I always found interesting is that often, both struggling companies and thriving companies compete with one another within the same market.

They face the same competition, have access to the same customer opportunities, and are impacted by the same local economy.

Why does one succeed and another struggle or fail? The foundational difference, I believe, is knowledge. Which brings me to Scott Ritchey.

I met Scott early in my career at a Carrier distributor meeting, and I knew immediately that there is something special about him. During our time together at those meetings over the years, along with countless phone calls brainstorming ideas or discussing day-to-day problems, I confirmed what I initially saw in Scott: He values knowledge and is constantly striving to gain more.

He is truly a student of small business. More importantly, he cares about people and is passionate about using his experience and knowledge to help them grow and succeed.

What Scott shares in this book is key to the success of any business.

While he has the knowledge and skills to have written numerous books about how to better market your company, improve your operational productivity or implement winning sales strategies, he chose to focus on the one thing that many small business-owners don't truly understand: Financial structure and the impact of pricing strategies. Scott's experience has shown him that a thorough understanding of these things most often leads to success.

Deciding to become a small business owner myself was not an easy decision. But it was a decision made easier because of the knowledge that I have gained over the years through my association with Scott.

Any success that I enjoy in my business will be owed, in part, to him. I've seen many people flourish under his leadership, and I'm proud to say I'm one of them. I consider him a mentor and I call him my friend.

When you finish this book, you'll understand what I have discovered: Scott Ritchey is the real deal. He knows what he is talking about, and he is passionate about sharing that knowledge. What you learn from Scott will make a difference in your small business. And it could be the difference between success and failure.

Pay attention.

Brad Johnson- Owner- AAA Northgate One Hour Heating & Air, LLC

Introduction

This isn't just another book about how to promote yourself on social media, get lots of HVAC jobs, and keep your existing customers satisfied. We know from experience that most HVAC business owners have plenty of work and most are well known in their local area unless they have just set up shop. And yet, too many HVAC businesses are falling by the wayside. It is a sad fact that 30–35% of HVAC ownerships fail by year 3, 47% by year 4, and a staggering 90% of them fail within 10 years. Why do most HVAC businesses fail? There's a single key reason: most HVAC contractors lack the business acumen needed to understand the financial ins and outs of running a successful business.

You may wonder who we are and why we take such a strong interest in contractor success in the HVAC industry. We aren't economists or business professors who are out to impress you with fancy language. In fact, we deliberately try to explain financial jargon in simple, easy-to-understand terms because we want all our readers to understand the important keys to success in this book. First, however, we would like to take a few minutes to introduce ourselves so you can get to know us and understand why we are approaching the HVAC failed business problem in this way.

I'm Scott Ritchey, vice president for the Plumbers Supply Company, writing with my colleague Gary Kerns, president and owner of Superior Heating and Air Conditioning, Inc. Our backgrounds have given us a unique outlook on the HVAC contracting industry.

I studied both accounting and business while as an undergrad at Pennsylvania State College, then earned a Master's in Business Administration (MBA). While working as a business consultant for an accounting software firm, I traveled to various cities coaching HVAC contractors on implementing our software, including how to analyze and interpret the financial data it provided. This led to my involvement with an HVAC company that later hired me as their general manager and managing partner to handle the company's business strategies.

As I officially entered the HVAC business with them, I soon discovered that the company was losing money despite taking in about $2 million in revenue each year. The reason was simple: the company just didn't understand that mark-up does not equal margin and that an HVAC business has varying overheads between its different departments. This lack of business knowledge showed up in their pricing methods, resulting in a 2% loss in the company's financial performance.

Worse, the company was making the fundamental mistake of thinking that all construction jobs were the same. Drastic change was needed to ensure that all jobs were competitively priced, yet would yield above-average net profits. With new visibility on each department, I helped the company see which types of construction and service work were the most profitable. By focusing on these more profitable areas, the company grew from $2 milllion per year to $6.5 million per year in a short five-year period before my interest in the company was sold.

Inspired to help other contractors suffering from the same lack of business acumen, I re-focused my career around providing sound financial training that will help any HVAC owner sustain a successful business. To date, I have worked with about 400 companies and thousands of individuals, enabling them to learn important financial principles that all but guarantee success in the HVAC contracting industry. I have nearly three decades of experience consulting with HVAC companies all over the United States, and the success rate of the companies I have worked with is simply astounding. Most remain successful long-term—and the ones that wind up closing do so after spending more than four years in business. In fact, I recently got a call from one of my clients telling me he was finally turning his business over to his son after 27 years. He thanked me for my help and let my know he never thought in his wildest dreams he would ever be worth $6.5 million. Now *that* is a success story!

Gary comes from an entirely different background. After high school, he joined the Air Force, serving as a mechanic for five years. However, he learned far more than just the basics of repairing jet engines. The Air Force imparted to him the importance of self-discipline, planning, and getting every single job done just right. In short, it gave him a passion for excellence that infuses every job he sets his hands to.

After his stint in the Air Force, Gary became an apprentice with a commercial HVAC company. It was a perfect fit. Gary's talent as a top-notch service mechanic came to the fore and he became enamored with the idea of starting his own HVAC company. After seven years of working as a company employee, he sought out a business professional in the HVAC trade and asked him the keys to success in the business. This professional responded by teaching Gary how to read a profit and loss statement, obtain and manage lines of credit, understand the details of a balance sheet, and treat employees well.

In 2001, Gary and I met at a dealer advisory council meeting—and the rest, as they say, is history. The result of this partnership is not just a successful business relationship but invaluable advice that can help anybody create a successful HVAC business as well.

This book is dedicated to every HVAC contractor who wants to run a successful business. Perhaps you have been a part of a failed past venture and you want to take another shot at ownership but aren't sure if you can make it this time. Maybe you've just started an HVAC company and want to ensure your new venture is profitable. Maybe you own a company that is not nearly as profitable as it should be and you can't figure out why. Alternatively, you may be an HVAC company employee who wants to start your own company but the dismal success rate for HVAC contractor companies in your area, coupled with horror stories of those who have tried and failed, scare you from making the leap.

This book can help you avoid the landmines that have tripped up many of the entrepreneurs who have come before you.

We don't claim to have all the solutions to every single challenge and problem you will ever face as an HVAC business contractor, but we do claim this: The business and financial information outlined in this book are guaranteed to work for any size contractor. Though some principles discussed in this book are static, the factors used in the principle can vary. For example, labor rates tend to be higher in the West and Northeast regions of the country in comparison to the Midwest and South. These differences could affect the outcomes of some of the formulas, making them more region-specific. Nonetheless, apply them and your company will turn a profit no matter what type of HVAC contracting work you do. Ignore them and you may fall become one of those 90% of HVAC companies that fails by year 10.

We want to help you avoid that by building a flourishing business you're proud of.

To your success!

Scott and Gary

Chapter 1
It's a Jungle Out There

We've all heard the phrase, "It's a jungle out there." It's commonly used by those who engage in shady business practices that benefit them while causing harm or pain to others. A high school or technical college teacher may have taught you the phrase to emphasize the need to stand on your own two feet instead of expecting the world to treat you with kid gloves. It is also frequently used in the business world to describe a competitive environment in which everyone is trying to carve out a small clientele as others attempt to take it away from them.

If you're approaching HVAC contracting with this mentality, it's time to stop. The following two analogies will give you a peek into what the playing field is really like in this line of business, enabling you to step back, de-stress, and refocus your efforts in the right direction.

The Lion and the Gazelle

The Lion and the Gazelle analogy is used in many motivation and leadership books. You have likely heard it before, perhaps retold in more than one way. If you haven't, the following version and its commonly accompanied application is shared by Christopher McDougall in his book *Born to Run: A Hidden Tribe, Super Athletes, and the Greatest Race the World Has Never Seen*:

Every morning in Africa, a gazelle wakes up. It knows it must outrun the fastest lion or it will be killed. Every morning in Africa, a lion wakes up. It knows it must run faster than the slowest gazelle or it will starve. It doesn't matter whether you're the lion or a gazelle—when the sun comes up, you'd better be running.

Another popular variation of this story was originally published in the science fiction magazine *Analog* but has been repeated ad nauseam both online and off:

A lion wakes up each morning thinking, "All I've got to do today is run faster than the slowest antelope."

Scott Ritchey with Gary Kerns

An antelope wakes up thinking, "All I've got to do today is run faster than the fastest lion."

A human wakes up thinking, "To hell with who's fastest, I'll outlast the bastards."

The problem with both these outlooks is that they are fundamentally flawed for one simple reason. While both animals are seeking to survive, the fact is that lions and antelopes/gazelles are fundamentally different. Lions are carnivores. They need to find meat to eat or they will starve. Lions may occasionally happen upon a wounded antelope and so be treated to an easy meal; however, in most instances, they must run—and run fast—in order to catch their prey.

Gazelles are herbivores. They eat grass, small bushes, short trees, and foliage. These food sources are abundant in the savanna, which means that gazelles never really worry about finding enough to eat. They don't need to run to find food. They simply need to wake up, take a look around, decide what to eat, and then start eating. If they don't like the taste of one little bit of foliage, they can simply walk to another patch and eat that.

When you look at the parable from a different light, it becomes apparent that it is a very accurate description of the HVAC contractor market today. Like the gazelle, HVAC contractors often find that there is food in abundance on the savanna. There are plenty of jobs to go around, so no one needs to face a shortage of work. Homeowners, home construction companies, commercial enterprises, and even government agencies all need HVAC services on a regular basis. In fact, there is so much work to go around that HVAC companies such as Gary's can be very successful long-term, even in small-town markets.

Unfortunately, many HVAC company owners make the deadly mistake of acting like the lion in the story. They fear that prey is limited and therefore pursue it with a passion. It's "kill or be killed" as far as they are concerned. Alternatively, they may act like the human in the second version and put a priority on outliving everyone else by offering rock-bottom prices.

The problem with acting like either the lion or the human is that in both instances, decisions are made based on a fear of failure. You will assume that you must do whatever it takes to get every single customer you possibly can. If you take this approach, you may fail to realize that your low proposal actually leads you to *spend* money on a job instead of making money from it. You may purchase low-quality supplies to cut costs, only to discover that doing so leads to unsatisfied customers and/or longer work hours as your crew attempts to get a job done without the proper tools. Another common mistake is to take on more jobs than you can realistically handle. In these and other similar scenarios, the results are never pretty. Fear of failure typically leads all of us to make decisions that we would never make if we felt our financial future was sound and stable.

HVAC does not necessarily need to be an overly competitive market. Sure, there are competitors out there—but there is plenty of work for everyone to be successful. You don't need to steal someone else's customers to gain ground in your business. There are plenty of niches, possibilities, and opportunities to go around. You don't need to hit the ground running; in fact, doing so is counterproductive unless you have thought long and hard about where you are running, why you are running there, and what you intend to do once you get there.

Sit back, take a deep breath, and let's move on to another analogy that can help you better understand the keys to success in the HVAC business.

Elmer Fudd and the FUD theory

In many of my financial seminars, I tell the story of Elmer Fudd as a light-hearted transitional story that lets me introduce a more pragmatic problem I see with struggling contractors: FUD. FUD stands for Fear, Uncertainty, and Doubt. I understand that Elmer was a hunter and you are a contractor, but bear with me, you'll soon see how FUD can prevent us all from seeing a clear path to success. It's a mental deterrent keeping us from taking chances or seeking help to navigate us through the journey called entrepreneurship.

As you recall, Elmer is one of the most recognized Looney Tunes characters. He is the arch-enemy of Bugs Bunny, arguably their most famous character. Elmer is passionate about hunting, specifically hunting rabbits. It's the same level of passion you may have for the HVAC or plumbing contracting business. But whether you are a hunter or a contractor, you can realize the fear of failure. If we were to watch an episode of Bugs Bunny vs. Elmer Fudd, we would clearly see how Fear, Uncertainty, and Doubt play out. Each episode starts out with a confident Elmer quieting the audience with his classic, "Be wery wery quiet, I'm hunting wabbits." But as the story unfolds, we see a highly frustrated Elmer whose fear of failure rises as he is thwarted by Bugs at every turn. Worse, he's frequently uncertain that Bugs is even a rabbit at all, as he is constantly fooled by the wascally wabbit's disguises. Both Elmer and the audience end up doubtful that he'll ever successfully shoot a rabbit.

Because FUD (unlike Elmer) is so deadly, it warrants a closer look. We'll see how it cripples HVAC contractors and learn what can be done to fight it.

Fear of Failure

The fear of failure is the deadliest element of FUD. Coupled with pride, it boxes many entrepreneurs into the corner of self-reliance, when obtaining outside help could make all the difference in the world. As the business begins to tank, entrepreneurs tend to become even more fearful. They don't want others to know about the company's pricing mechanics, operational costs, or overall profitability. They don't want to take on employees or outside vendors and consultants who could help turn the company around, because doing so would involve divulging company secrets to outside parties, exposing the failing condition of the company or the owner's lack of business sophistication.

You can't get help if you are afraid of making mistakes. Fear will hold you back from learning new skills that could benefit your business. It will hold you back from bringing people on board who could help counter and support your weak areas. It might also hold you back from branching out into a specific niche market that could wind up being extremely profitable for your company.

This theory is supported by a 2005 article from the Service RoundTable Organization, "Top 21 Reasons for HVAC and Plumbing Business Failures" Failure to seek advice or assistance from outside resources was the number two reason why these businesses fail. Very few owners starting a contracting business have a thorough understanding of everything it takes to be successful. If they waited until they knew everything, they would never get started.

If you find yourself lacking in a certain area, don't avoid the matter, hoping the business will survive despite this lack of understanding. Remember the weakest link analogy. Don't let the weakest link break a chain that is strong in other areas.

The Solution

The key to getting rid of fear of failure is to do what you are afraid of. If your balance sheet terrifies you, show it to an accountant. If you are afraid to learn how to handle your company's finances, either hire someone to do it or start learning the basics yourself. If you fear handling a certain type of project, learn all you can about it and then try it out. You conquer fear by doing what you're afraid of and then discovering that in most cases, it wasn't nearly as bad as you thought. At best, you could make huge profits. At worst, you will learn some important lessons that will help you be successful next time around.

Uncertainty

Uncertainty is the leading cause of impaired judgment and poor decision making. In Elmer's case, he was never quite sure if Bugs Bunny was a rabbit or not. This uncertainty kept Elmer from achieving his goal of shooting the rabbit even when Bugs was standing right in front of him. Uncertainty or lack of business experience is the cancer that slowly eats away at a contractor's success in managing his or her company's finances.

This is a hard one for most contractors to accept. Many contractors are afraid to share their uncertainty or lack of business acumen for fear of exposure or embarrassment.

This may come across as a harsh statement, but consider the following fact: HVAC businesses are more likely to fail than any other startup. In fact, only new restaurant owners have a harder time building a successful business than HVAC contractors.

The Solution

Naturally, there are some things that you simply do not know or cannot predict. However, you should never simply accept uncertainty as part of your business unless it is totally unavoidable. Financial uncertainty is particularly deadly; in fact, it probably causes more business failures than just about anything else. You need to know your company's financial standing at all times. This includes not just how much materials cost, your gross profits, and your competitor's rates but also the cost to profit margin relationships, tax rates, and much more.

Thankfully, you don't need to be a financial wizard to learn how to manage your company's finances. The simple fact that you are an accomplished HVAC professional shows that you have the smarts needed to handle business finances. Step out of your uncertainty and learn the skills needed to properly estimate how much you should charge for each type of job. The difference this simple step makes to your business will likely surprise you.

Doubt

Doubt creeps in when decisions have been made but the outcomes are not what you anticipated. It is usually at this point that the damage caused by uncertainty leads to failure. Elmer's realization of doubt happens at the moment he figures out Bugs is indeed a rabbit but has just made his getaway; Elmer frantically chases after him, shooting randomly. That moment of doubt happens to contractors between years 3 and 4. This is often the point when a contractor who has been very busy simply cannot believe he owes his suppliers a large sum of money and does not have the cash or accounts receivable to pay his overhead plus accounts payable. Unfortunately, like Elmer, he has been shooting at jobs with random prices that often were not profitable.

The Solution

Doubt, like uncertainty, can be overcome by certainty. Take control of your life by owning your mistakes and looking for ways to rectify them. If your suppliers are knocking at your door claiming you owe them large sums of money, investigate the claims. If they are true and you don't have the cash on hand, work things out with them and then assess your business to discover why was is a shortfall. In most cases, you will probably find that you are not charging as much as you should; however, don't automatically assume that is the case. Maybe you are also buying supplies you don't need or purchasing more than you need. Maybe you are paying more taxes than you should; in such a case, professional help from an accountant can help you reduce costs.

Never let doubt get in the way of your success. Don't doubt yourself and your ability to create a successful business. Don't doubt your employees and their ability to help you build a successful company. Have faith that even a bleak outlook can be turned around if you learn the right skills and implement them. In the worst-case scenario, you may need to close and start afresh—but even that is not necessarily bad. If you've gotten this far, you have what it takes to make a new start, and the lessons you've learned from your old business will create the foundation for success in your new venture. Remember: Walt Disney filed for bankruptcy twice before "Disney" became a household name worldwide.

Everyone Is Busy Every Single Day, and That's a Problem

The HVAC contractors I have met are extremely hard-working individuals. They worked hard when employed by a company and they worked even harder to start their own business. But as it turns out, hard work alone does not create a successful business. There is a common—yet completely wrong—misconception that you can build a successful HVAC business simply by providing quality service to customers at a good price. Nothing could be further from the truth. While good service and good products certainly play a role in every company's success, the fact is that alone, they won't take you very far.

Strategic thinking is critical for any business owner to be successful. A business owner must take the time to get a big-picture perspective, analyzing each component of the business and how it intertwines with the other components. A wise entrepreneur will do this even before starting a new business—but it's never too late to start. In fact, getting a big-picture perspective is something that should be done on a very regular basis; otherwise, things will run on autopilot and most likely crash at some point in time.

What should you look at when you want to see the big picture? Well, the simple answer is to look at *everything*. Literally every single detail of your business should be examined to determine if something could be made more profitable, more efficient, or more effective. Look at your suppliers to see if they are offering you the best products at the right prices. Look at your workers to see if they are the people you need for the job at hand. Look at how work is done to see if there is a better and/or faster way to do things. Pay attention to small expenses, small supplies, small tools, and small employee habits. Every little detail will affect your HVAC business both now and in the future.

Even more importantly, look at the direction your business is headed and see if it is indeed the best direction for you. There are a lot of niche fields in the HVAC industry. Some companies handle all sorts of jobs, including residential replacement work, new home construction work, and commercial HVAC work. Some specialize in a single type of service such as maintenance and repair, or HVAC installation and replacement. Some companies only handle jobs in their locality, while others are willing to send workers to other cities and towns.

There is no one right model for all HVAC companies. A business owner will need to analyze supply and demand, look at his or her workers' skill sets and experiences, and, even more importantly, consider the profit margin in each field to decide what type of jobs the company should accept or focus on pursuing.

As we noted above, the "It's a jungle out there" mentality leads business owners to make decisions they would not make if they were thinking with a long-term vision. The jungle mentality is a narrow-minded, short-term way of looking at things that will never lead to long-term success. Jungle animals do not think ahead; they do not plan out their entire day, much less envision weeks, months, or years in advance. Their only goal is to find food and water that day at any cost. That is no way to run a successful business. Unlike jungle animals, an HVAC contractor needs to invest in certain areas of overhead to (hopefully) earn money in profits.

If the key objective of any business is to be profitable, then all things related to profitability must be examined to maximize that goal. Being a successful business owner requires pig-headed discipline to understand and pay attention to the thing that matters most: the bottom line. Unfortunately, lack of business acumen is a key factor that keeps too many business owners from being able to maintain this strict focus.

Chapter Summary

There are many rewarding reasons to become an HVAC entrepreneur. Though making more money is very attractive, helping people solve their HVAC problems and creating jobs for others provides even greater satisfaction. It is important to realize that you do not have to go it alone. Use the help of others like your suppliers, accountant, or banker to help you navigate your journey—especially if you lack the business acumen necessary to run a successful contracting business. Do not allow FUD to keep you from being successful. Remember: there is plenty of work to go around, so you want to seek out the most profitable jobs to ensure your company's financial health. Financial health starts with understanding the basics of pricing for profit and controlling your company's costs, which is the focus of Chapter 2.

Chapter 2
Profits through Market Cap Pricing

Chapter Overview

In this chapter, you will learn:

- Pricing methods and why they are important to your business
- The difference between gross profit and net profit
- How to understand market caps for each business segment
- How to calculate the right gross margin for your overhead
- Markup does not equal margin
- How to calculate monthly breakeven sales

Fair prices do not translate into good profits. I have met many contractors who offered a fair price but who still lost everything and wound up having to work for someone else.

Why? The problem is that "fair" is not an objective standard. After all, what *is* fair, anyway? Is a fair deal one that is great for your customers? Is it fair because you feel good about it or another person feels good about it?

Many HVAC contractors define a fair price as one that is comparative to the price being offered by other contractors for the same type of job. This would seem like a good standard to adhere to—but it's not. After all, not all contractors offer the same high standard of service. Another company may be able to offer a better deal because most of its employees are in fact apprentices who are being paid less than skilled licensed journeymen and are not earning benefits. Alternatively, another company may be able to charge less because it is not required to complete the job within a short period of time and/or is using low-quality materials.

Even more importantly, stop to consider the fact that nearly 47% of contractors do not survive their fourth year in business. If you are basing *your* prices on *their* prices, you are setting yourself up to lose. Why? The sad and simple reason for this problem is that most contractors know how to do a job, but they don't know how to accurately determine how much net profit they are making from the job.

I have been teaching financial management classes for 27 years and see the mismanagement of pricing policy every time I run a class. In fact, when I give 10 contractors in a class the same job cost and overhead and ask them to make a 10% net profit, it's quite common for me to see 10 different prices set! This concerns me greatly, and I am convinced this is a leading contributor to contracting business failures.

For fun, let's do a self-test right now. In this example, your cost to do the job is $2,700, which includes all equipment and materials, labor, tax, and permits. You have a 38% overhead and you want to make a 10% net profit on this job. What is the correct price for this job to yield the 10% net profit you desire?

We will compare your answer with the correct price later in the chapter. No peeking ahead! A true comparison will provide great insight into how you may be doing things now and how our method does it.

Not understanding the relationship between costs and pricing is the main reason for lack of profitability. But this is just one factor. To maximize profits, a contractor must understand all the components that make up profitability. These components come under the two headings of price control and cost control. We'll start with price control and then get to know cost control and all that it entails.

Market Segments Dictate Margin Caps

When I say "margin caps," I am referring to the top price a customer is willing to pay for a product or service. I am also taking into consideration the fact that many HVAC businesses have several target audiences or markets they are reaching out to simultaneously. These markets could include a residential new construction business, a commercial new construction business, a residential existing home replacement business, a commercial existing building replacement business, a residential service business, and a commercial service business. To complicate matters, residential new construction can be broken down into tract or custom homes, while commercial new construction can include plan and spec construction or design build construction. Each of these segments has its own market margin cap that affects top market pricing.

Most contractors I have talked to are unable to give me the correct answer when I ask what the margin cap is for any one of these business segments. These market margin caps reflect the maximum amount of gross margin that a price will yield in each of these segments.

The table below lists the top market margin caps for each business segment in HVAC.

MARKET MARGIN CAPS BY BUSINESS SEGMENT	TOP MARGIN %
Service Repair Sales	70%
Existing Home Replacement Equipment Sales	50%
Commercial Equipment Sales	40%
New Construction Custom Homes	38%
New Construction Tract Homes	32%

Years ago, I had an interesting conversation with a contractor in South Carolina named Mr. Lambright. We met up after one of my financial management classes to discuss my thoughts on the market cap for existing home HVAC replacement services in his town. He found it hard to trust my numbers, so I encouraged him to try an experiment. I told him to keep adding $100 to his HVAC replacement bid proposals until he lost a job for charging too much money.

The next time I saw Mr. Lambright was at a fall training class on marketing. After the class, I noticed him hanging around as the room was emptying out. I asked if he had any questions; he said no, but he wanted to talk about our conversation from the spring when I encouraged him to keep adding $100 to his quotes until a customer told him that his price was too high. When I asked him how the experiment worked out, he began laughing. He said, "Mr. Ritchey, I had to add $100 10 times until someone told me my price was too high."

Take a moment to digest that number. Every time Mr. Lambright took on a job, he was charging $1,000 less than he could have been, simply because his prices were $1,000 below the market cap. As we discussed his new job proposals and the local market cap in depth, he realized that the amount he was now charging for his work was the market cap that I encouraged all seminar attendees to charge for their services.

Mr. Lambright is far from being alone. Literally tens of thousands of contractors around the country are pricing jobs $1,000 to $1,500 below the top market price. This was substantiated in a 2010 survey conducted by Decision Analytics for Honeywell. In that survey, homeowners said they would pay up to $1,500 more for a system that they felt delivered on the expectations they had for that system.

Some may discard this survey as being dated, but recent data backs it up. Surveys and studies done by Unico and Emerson Climate Technologies show that homeowners are less concerned about the price tag and more concerned about comfort, energy efficiency, air filtering and air quality, and the warranty provided on the new system. If you are offering a product that meets customers' high standards and expectations, don't be afraid to charge the market cap price for your goods and services. If a customer balks at your price, don't immediately back down as if you have done something wrong. Instead, explain why your product and services are worth that price. Chances are, customers will be happy to pay what you are asking when they understand what they are getting in return.

Another interesting nugget from the Honeywell study I mentioned above is that 62% of homeowners indicated they were ready to make their purchase immediately. Yet this isn't what most HVAC businesses do! Up to 77% of contractors do not close the sale during their first conversation with a homeowner, opting instead to draw up a written proposal and send it for consideration. This practice looks great on paper but does not meet the need of the 62% of customers who wanted to buy right then—which opens the door for that homeowner to get another bid, thus ruining your chances of landing the job.

If you are one of the many contractors who opt to send a written proposal instead of giving an immediate price quote and asking for the job, I urge you to stop. You are acting like your own worst enemy in these situations. We talked about stepping out of the jungle mentality and stopping to think, consider, and plan in the first chapter of this book—but now is the time to stop thinking and planning and start picking the low-hanging fruit. If a customer shows an interest in your goods, services, and price quote, take the job. It is yours to lose. Rest assured that a customer who wants a price quote and other bids from various contractors will let you know.

What Is Profit?

This is a great question, but before I can answer it, we should get greater insight into what it's really asking. To help, let's look at a basic snapshot of a profit and loss statement, also called an income statement.

A basic profit and loss statement, or P&L, has five sections. Simply put, it tracks the flow of money through the company. Every dollar generated from a sale ends up in the "revenue" section. As you pay for equipment, materials, labor, and miscellaneous expenses directly related to the job, this distribution of cash shows up in the "cost of sales" section. The dollars left over are the gross profit dollars that are used to pay overhead expenses. After the overhead is paid, the remaining dollars are your net profit. These are the dollars you get to keep as personal income or reinvest in your business.

Depending on your experience with profit and loss statements, this explanation was either very basic or immensely helpful. But understanding the P&L is not the point I want to drive home here. As you can see from the P&L below, the word "profit" appears twice—and this is where I often find confusion when speaking to contractors about the profit they made on a job.

Basic Profit & Loss Statements

Revenues	$1,000,000
Cost of Sales	$600,000
Gross Profit	$400,000
Overhead	$350,000
Net Profit	$50,000

I have spoken with numerous contractors who have bragged about winning a job and earning a certain amount of profit from their work. My response has always been, "Is that enough?" Naturally, contractors who hear that question get perplexed. They ask, "What do you mean?" I then rephrase the question and ask them if the money they claim to have earned in profit was the money left after all expenses had been covered. In most cases, the contractor I am talking to will affirm that the money he or she considers to be profit is indeed the money left after all expenses are taken care of. Once again, I ask the question, "Is it enough?" and once again I get perplexed looks.

You see, the profit the contractor is referring to is the *gross* profit, which is not *net* profit. That is when I ask the contractor to tell me what the overhead was on the job in question. As it turns out, most contractors cannot answer that question. They simply do not know what the overhead was on the job they just did—which means they do not actually know if they made a lot of money, a little money, or no money at all.

The most common mistake a contractor makes in pricing a job is believing that margin is the same as markup when this is not actually the case. The idea that margin and markup are two separate things is not new; in fact, I first became acquainted with this concept when I attended a Stu Doctor financial class for HVAC contractors in the early 1980s. Stu was a renowned financial instructor for the Carrier Corporation who taught HVAC contractors financial principals much like I do today.

A very common pricing practice used by contractors is the multiplier method. They will take the cost of the job—which usually includes materials, equipment, and labor—and multiply it (markup) by some factor that they believe to be a margin on the job. There is nothing wrong with the markup method if you know what the multiplier needs to be to produce the correct margin you want to achieve.

This, however, is exactly the problem. Most contractors do not know what the multiplier should be, which is why they make the "markup equals margin" mistake and incorrectly price the job.

For example, a contractor may want to make a 35% gross margin on a job, so he will take the cost of the job and multiply that by 1.35, thinking he will make a 35% gross margin. Unfortunately for the contractor, the result is *not* a 35% gross margin. Let's look at the math.

Let's say the job cost, which includes equipment, materials, labor and any related permits, reserved warranty cost, tool rental, or subcontract labor like an electrician or non-employee labor, amounts to $3,200. $3,200 times 1.35 comes to a sale price of $4,320, so the contractor submits this price to the customer with the expectation of making a gross profit of $1,512, which would be 35% of the sale price of $4,320.

Unfortunately, the math does not add up as the contractor expects. When we take the sales price of $4,320 and subtract the job cost of $3,200, we come up with a gross profit of $1,120, nearly $400 or 9% less than the contractor expects.

This shortage represents the difference between a profitable contractor and a struggling one. My experience as a business coach and HVAC company consultant has shown me time and again that failure to properly calculate markup and margin is the biggest and most concerning problem affecting contractor profitability and a contractor's chance for survival as a business owner.

Know Your Breakeven

If you could predict each month what your company must do in sales volume just to pay all the bills, would you relax a little more knowing that your business is at least breaking even?

Well, you can—and it is an easy formula. Let's say your monthly expenses excluding direct labor, equipment, materials, and permits are $10,000 per month. These expenses are there whether you have any work or not. Now what? Knowing the gross margin percentage of all your business segments combined will allow you to determine your breakeven sales number to recover that $10,000 monthly overhead.

For example, let's say my business concentrates on the existing home replacement market and I sell my jobs at a 45% gross profit. I take my $10,000 overhead and divide it by the 45% gross margin to find that my monthly sales breakeven is $22,222 per month. Seeing this, I know that for every sales dollar over that $22,222, I will make a positive net profit. I also know that at $22,222, my overhead is 45% ($10,000/$22,222). The average replacement sale is around $6,500 to $7,000 for standard 14-SEER equipment. If I divide my breakeven sales of $22,222 by the $6,500 average price, I find I need to sell 3.5 replacement jobs per month at a 45% gross margin to break even. As you can see, knowing the breakeven shows me the minimum number of jobs I need to do each month just to pay the bills.

If you have been in business for a few years or more, it is much easier to calculate your overhead percentage. Keeping that $10,000 figure as the monthly overhead, your annual overhead would be $120,000. Now look at your sales figures for that same 12-month period—for this example, let's say the sales revenues were $350,000. To find your overhead percentage, simply divide that $120,000 overhead by your

$350,000 sales revenues: the overhead percentage is 34% ($120,000/$350,000 = 34%).

Knowing your overhead percentage is critical to pricing jobs for positive profits. Your calculations will be far more accurate than before, enabling you to turn a profit on each job you take on. Remember, every sales dollar above the breakeven sales number will start producing net profits at the rate of the gross margin percentage.

Do You Want to Earn a Good Profit on Each Job?

Now that you understand what you need to charge customers to earn a profit, a single question remains: Do you really want to earn a good profit on each job? Your automatic answer may be yes, but take some time to consider your current business model before assuming this is a dumb question. Before I get into the statistic of how many people truly base all buying decisions on price, let's solve the practice problem I set up at the beginning of the chapter to get another look at the single-factor pricing method.

In our self-test example, your cost to do the job, including all equipment and materials, labor, tax, and permits, was $2,700. You had a 38% percent overhead and wanted to make a 10% net profit on the job. What is the correct price for this job to yield the 10% net profit you desire? Let's solve it now.

To calculate the right price, the first step is to add your 38% overhead plus the 10% desired net profit. This yields 48%. This 48% represents the gross margin percentage you have to make to pay the 38% overhead and leave you with the 10% net profit.

Since the single-factor method divides the cost of the job by the inverse of the gross margin percentage, we must find out what this inverse is. To do this, take 100% minus the gross margin percentage needed. In our problem, 100% - 48% (gross margin) leaves you with the inverse (better known as the divisor), which in this case is 100% - 48%, or 52%.

Using the single-factor pricing method, we can now calculate the correct price to yield a 48% gross margin. Simply divide the $2,700 by 0.52 (52%) and you get a sales price of $5,192.30 for this job.

Now let's prove the formula. Take our sales price and subtract our job cost from it. In this scenario, $5,192.30 - $2,700 equals a gross profit of $2,492.30. This is gross profit, not net profit. To find our net profit, we take the gross profit we just calculated and subtract our overhead for the job.

The question is, how much overhead did we have on the job? Remember our overhead was 38%. Knowing this, we take the 38% and multiply it against our sales price. So $5,192.30 x 0.38 = $1,973.07. This number represents the overhead figure on this job.

Now the only thing left to find out is the net profit number left over after the overhead is taken out and whether that equals a 10% net profit. To do this, take the gross profit number and subtract the overhead number, leaving the net profit number. In this example, we get $2,492.30 - $1,973.07 = $519.23 net profit. Dividing the net profit by the sales price gives us the net profit percentage: 519.23 / $5,192.30 = 10%.

As you can see, the math adds up and the formula is proven. Once you trust the math of the single-factor pricing method, you will deliver the net profits your company deserves.

The Low-Price Boogieman Is Not That Big

The reason I asked whether you always want to make a profit is because it's really a way to find out whether you believe the lowest price is what always gets the job. Industry statistics show that up to 30% of consumers shop on price alone. What this means to you is that 3 out of 10 potential customers who ask you for a quote do not care about comfort, the quality of your products, or long-term savings. They simply want the best possible price. In contrast, a full 7 out of 10 people want a value proposition—they want to know that the price matches the value they think they are getting. To them, it is not all about price.

So the question is, who do you want to sell to and make your lifetime customer? If you choose to sell the price shopper category of customer, be prepared to enter a bidding war. More importantly, understand whether you can afford to be in such a war.

What happens when you find yourself in the middle of a bidding war? You know it's going to get ugly and if you really want to go there, you must decide what is your threshold before you fold 'em. Sometimes you may succeed and be able to drop to the lowest denominator to win the job, but in many cases, it isn't a good idea.

Let's consider the following scenario. You are bidding for a job. You have estimated your overhead costs, the cost of doing the work, and the net profit you need to earn to keep your company in business. Your bid for the job comes to $4,400. However, a few days later, you discover that a competitor has bid for the same job and his or her price tag is $3,900. The customer says, "I like you and want you to do the work, but you have to match your competitor's price or you can't have the job." What do you do?

What *should* happen is that you take a second look at your proposal because you might have some wiggle room. Perhaps you can afford to shave $500 off your price and still manage a positive net profit. If this is the case, then consider matching the competitor's proposal, especially if the job offers the possibility of additional work in the future. Remember, the only way you will know if you are still making a profit is to look at the gross margin dollars you anticipate making on the job and subtract the overhead for the job. If you still have a positive net profit after that, you can reduce the price by $500.

At the same time, bear in mind that there are cases when you simply cannot match a competitor's proposal and still realistically turn a profit on the job in question. You can try to reduce costs, but reality is reality. If you cannot make money on the project, you may need to turn it down. This could bother you or make you feel guilty about turning down work over a price tag, and there is a chance that the potential customer will feel upset that you are not willing to do the job at the same price as your competitor.

All isn't lost, though. In some instances, you may be able to point out that your goods and services are superior. Perhaps you can get the job done faster than your competitor. Maybe your HVAC units are more energy-efficient than a competitor and/or have a better warranty policy.

However, at the end of the day, you may simply need to walk away if you cannot win the price war. Don't feel bad; it happens to every successful contractor and will most likely happen to you more than once during your time in business.

When you feel an intense desire to do whatever it takes to bid lower than any other HVAC company in the hopes of getting a coveted job, step back and remember that while 30% of consumers consider price to be the most important factor when choosing an HVAC company, the other 70% of consumers make HVAC-related decisions based on considerations such as long-term savings, product warranties, comfort, and air quality. These are the people you want to reach—and thankfully, there are plenty of them out there. Do you really want to stick to working with the 30% who continually demand lower prices, or would you like to work with the 70% who truly appreciate good-quality products and great customer service, even if it means they need to pay a bit more to benefit from all you have to offer?

Don't forget the lion and gazelle analogy. Every day, contractors awake to a field of opportunities. You can be busy doing work for the 70% of consumers willing to pay your price or get so busy working with the 30% of consumers who want to pinch every penny that you starve to death from low to no net profits. The choice is yours.

Chapter Summary

There is a saying in the sales industry that pricing is a contact sport. That is a very true statement for the HVAC field. Because so few contractors understand the basics of pricing mechanics, we see too many different prices for the same job. From this chapter, you should now understand these mechanics and how they relate to profitability. Knowing market margin caps and that margin does not equal markup are key cornerstones to building a profitable foundation for your business. Mastering the single-factor pricing method will strengthen your financial foundation, allowing you to add the pillars of growth and market expansion for your company.

Scott Ritchey with Gary Kerns

Chapter 3
Cost Control: Managing Your Overhead

Chapter Overview

In this chapter, you will learn:

- The income statement structure
- The basic definition of overhead
- The six categories of overhead
- How to benchmark the six categories as a percentage of sales
- Ways to control overhead

Now that we understand how to manage pricing for segmented market caps and how to accurately calculate a profit margin for every single job you take on, it's time to dive into managing overhead. Put in simple terms, this means that you need to understand and manage your expenses so you can ensure they are covered when you take on a job.

Why is this important? The better you manage your overhead costs, the more competitive you can be in your pricing and still make a good profit.

How Reducing Overhead Impacts Your Prices

Have you ever heard the phrase, "Sounds easy, does hard?" Reducing overhead is something that sounds easy, but as you dig into it, does hard because you must make some tough choices. Just remember to ask yourself: Does that expense have a real purpose? Most of the money-saving recommendations I make won't save you large sums of cash, but every little bit helps. Shaving a mere 1–3% off your overhead can make a difference in how you bid and secure work.

Let's say, for example, that you put the tips we share later in the chapter into practice and you manage to reduce your overhead by up to 3%. That doesn't seem like much, but it is a step in the right direction to providing a customer with a more competitive price without sacrificing net profit, if that is what you need. Another option would be to allow these cost savings to drop straight to the bottom line, resulting in higher net profits.

Let's put this in practical terms once again: Suppose the ABC Company has an annual overhead of 40%. They are about to bid a job that has a cost to do the work of $3,200, which includes all labor, equipment, materials, and permits. ABC likes to make a 10% net profit on all their system replacement work. Knowing this, ABC would have to sell that job for $6,400 to make the 10% net profit.

Now, if the ABC company put the cost savings we'll share into practice and reduced their overhead by 3% to a total of 37%, they could bid the same job with a cost of $3,200 and the same 10% net profit for a new price of $6,037, with a savings of $363 to the consumer. This allows the ABC Company to be more price competitive without sacrificing their net profit percentage—there's only a $37 difference in net profit dollars but an improved chance at securing the job.

On the flipside, If the $6,400 price was competitive enough to secure the job and ABC reduced their overhead by 3%, they would now have a 13% net profit instead of a 10% net profit. In net profit dollars, they go from $640 to $832. That is a 30% increase in net profit dollars: $832 − $640 = $192 / $640 = 30%.

Being able to increase profits allows you to upgrade equipment, offer your workers an annual bonus, or simply enhance the bottom line. Remember, you don't have to lower your prices every single time you are able to do so. If you do good work and know where to go to find jobs, you can charge a reasonable price for your services without worrying that you will lose customers.

As I stated before, when I do a financial workshop, 9 out of 10 contractors typically don't know how to accurately calculate their overhead to ensure they are not losing money on jobs. This has always troubled me greatly, since the situation was always the same no matter where I taught or how long contractors had been working in the HVAC industry. As I dug deeper into this lack of understanding of the role overhead plays in pricing success, I discovered a low level of knowledge about the structure of the income statement.

Suddenly, a light came on—I realized that this condition existed because of just one or two factors, and quite often both. The first root cause was contractor FUD. The reason many contractors in my classes did not know the difference between markup and margin was because they never asked for help from anybody before attending one of my seminars—they feared exposing their lack of financial business acumen. Their lack of basic accounting knowledge was contributing to their financial struggles. The second root cause I discovered was that distributor sales reps did not possess the business acumen necessary to teach their contractor clients these financial principles, thus also contributing to the contractor's struggles.

I can honestly say that the contractors who have taken my classes in the past and understand the method I use to price a job are either still in business and very profitable, or at least lasted longer than the four-year average in business. The ones who failed after lasting longer than that four years typically lost their businesses because they did not understand the two most important income statement sections, which control not only profitability but also market competitiveness. These sections are the labor portion of the cost of goods sold section and the overhead expense section. Contractors who fail in this area simply do not understand what these sections are telling them or they mismanage these expenses.

Let's start by taking a close look at the income statement structure, then dive into our discussion of overhead and the role it plays in pricing.

Income Statement Structure

The income statement, which is often referred to as the profit and loss (P&L) statement, accounts for where all the sales dollars go when paying the direct cost on the job, such as equipment, materials, and the labor to do the work. The remaining sales dollars left after the cost of goods are paid are the gross margin dollars used to pay the overhead items. Any sales dollars remaining after the overhead is paid become our net profit.

A simple illustration of an income statement is the revenue pie chart on the next page. The total pie represents a typical contractor with $500,000 in annual revenues. Though not exact, these numbers are not far off from what we typically see when we do financial consulting work, plus or minus 3-5 percentage points. The good news, as you will see in future chapters, is that we have tools that can double and triple the net profit of 5% used in the example. We will do a deep dive into the value of the income statement and what it tells us later in the book. For now, we want to focus on the overhead section of the income statement.

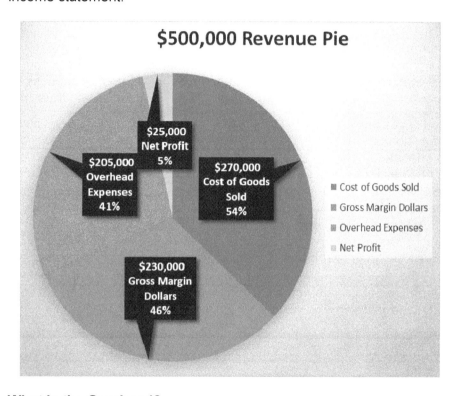

What Is the Overhead?

The basic definition of overhead is the expenses that your business must pay whether you have revenue coming in or not. It is the most important factor in deriving the right price on a job to ensure you make a profit. After all, no one would purposely price a job to lose money. Right?

Well, to avoid losing money, our pricing formula must include a way to account for the overhead that would be assigned to that job. To do this correctly, you must know your company's overhead percentage as it relates to your revenues generated from sales dollars.

Try making a quick list of overhead items—you should have expenses in most of these areas. These items include owner's wages, office wages, payroll taxes, insurance expenses, fuel and truck repair expenses, rent, utilities, legal and accounting fees, uniforms and small tools, advertising, telephones, computers, internet services, training, dues and subscriptions, bad debts, interest expense, office supplies, unapplied labor, and depreciation, just to name a few.

If you don't get an income statement prepared by an accountant or software program, the best way to figure your company's overhead is to add up all your monthly expenses (bills) that you must pay regardless of whether you have any work going on. Take that monthly number and multiply it by 12 to get your annual overhead expenses. Hopefully you keep a log of all the sales you did for the year to find your annual revenues. You then take your annual expenses number and divide that by the annual sales number—this gives you your overhead percentage.

At this point, I do want to caution you about hidden expenses—especially if you are a new startup company or if you set up a shop at your house. Hidden expenses are more relevant in these cases because you may hide or not account for all business expenses, instead lumping them in with personal expenses. For example, I often see contractors who work from home, but don't count their home mortgage and the associated utilities as business overhead. Why not? If they were in a rented office, they would count it as overhead in the form of rent. This is just one area where I typically see contractors misrepresent their overhead and fail to account for it properly.

As we get into the different categories of overhead, I want to make a significant point. Any overhead item must have a purpose, and that purpose is to support the sales and/or operations of the business. If it does not do this, then it is considered waste, and waste does nothing but eat up profits unnecessarily. To control overhead costs, you need to eliminate waste from your overhead. An example could be unapplied labor. In our industry, I have seen reports that suggest that up to 30% of a service technician's time is unapplied time due to inefficient dispatching or scheduling. There are other examples of waste, too—you'll need to be aware of your own business's needs. I just want to make you aware of this issue and how it impacts your overhead percentage.

There many items that can be overhead expenses, but I like to simplify it by categorizing these expenses into six broad categories and then analyzing them to see how they impact your business as a percentage of sales profits. I call these six categories "the stairway to total overhead." I have placed average values in the graphic below, but keep in mind these can vary +/- 2–3%.

Scott Ritchey with Gary Kerns

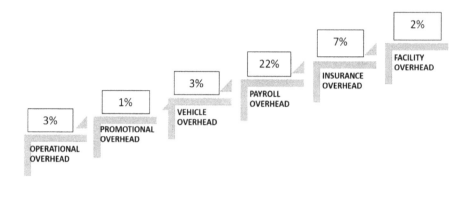

Average $500K–$1.5M HVAC Contractor's Overhead Range is 37% to 40%

Operational Overhead

Operational overhead expenses usually account for up to 3% of sales revenue. They include one-time expenses such as obtaining a business license to operate in the market you serve, retaining an accountant and/or an attorney to help you set up as an LLC or S-corporation, and purchasing a computer and accounting software like QuickBooks to keep track of business transactions. This category of expenses also includes office supplies such as business cards, stationery, and invoices with your company logo. It includes a landline, a fax machine, a printer, and a photocopier if you think you really need one. Ideally, you will want to keep these costs at 2.5% to 3%.

Promotional Overhead

Startup companies naturally need to spend more on promotion than established companies. However, even companies that have been around for several years or longer need to spend money on advertising and promotion. I understand referrals play a big part in generating revenue for your business and you probably prefer it that way, but the truth is, even a person with a large network cannot survive on the business of his or her network alone. If you are a growth-minded contractor, you are going to have to make investments in some form of marketing vehicle and so you must know the impact of this expense and how it affects your total overhead.

MAKE MORE MONEY: 12 Profit Pillars For HVAC Contractor Success

In general, industry statistics show that HVAC companies spend about 1% of sales on promotion. This is far less than what *should* be spent if you want to be successful long-term. In fact, evidence from many studies clearly indicate that an HVAC company should spend up to 5% of its overhead budget on advertising and promotion to enable sustainable growth.

Vehicle Overhead

There is no asset, other than human capital, more important than the vehicles needed to take your business to your customer's home or office. Although only the interest portion of a truck loan will appear as an expense on your income statement, there are many costs associated with company vehicles. For most contractors, it comes as a surprise that truck maintenance and fuel eats up about 3% of their revenue dollars. You could cut maintenance expenses by avoiding regular maintenance check-ups, but doing so is tantamount to shooting yourself in the foot, because it typically costs far more to replace a damaged part than it does to tune up the vehicle so it continues to work properly.

Payroll Expense

Human capital is not only the largest expense in any HVAC business but also the one that is most likely to get out of control. When I review a contractor's financial data, I often see a common mistake in how the profit and loss statement (income statement) is set up. Many times, direct labor and overhead labor are combined and listed in the overhead section of the income statement as wages. Treating all wages as overhead overstates what your true overhead percentage should be and could cause you to overprice a job quote without realizing it.

Direct labor—that is, labor needed to perform a specific type of work on a job—must appear in the "cost of sales" section of the income statement. Support labor or overhead labor is always present whether you have revenue producing jobs or not. Put in simple terms, you must pay support labor an hourly wage no matter how much or how little work you have. Indirect labor costs such as office wages or managers should appear in the overhead section of the income statement. Industry average overhead payroll is about 20–22% of sales revenue.

27

Understanding industry benchmarks for labor control always makes contractors more profitable. These benchmarks include labor as a percentage of replacement sales, labor as a percentage of service revenue, staffing, throughput ratios, and payroll compensation ratios. Don't worry—these concepts are not as complicated as they sound and I will be discussing them in depth in the benchmarking section of the book.

Insurance Overhead

No company owner enjoys paying insurance, but it's a cost that you should never consider skimping on. If you don't have the insurance you need, your company is at grave risk if something catastrophic happens. To be safe, you should probably assume that something catastrophic will happen at some point in time, especially if you run a successful HVAC contracting company long-term. Employment estimates from the Bureau of Labor Statistics show that HVAC work is one of the top 10 most dangerous jobs in the United States. The work becomes even more dangerous if, like many HVAC companies, you take jobs working at construction sites.

You need at least the basic insurance required by your state. Liability insurance is a must so that customers will not sue you if a portion of their building is damaged or if they sustain an injury because of your work. Workers' compensation is needed to provide medical care and financial assistance to employees who are injured while working for you. This insurance provides payment no matter who was at fault for the injury. Casualty insurance provides for the family of a worker who dies on the job.

If you have a large business office and you store a lot of expensive equipment on-site, you will probably want insurance coverage that will provide compensation in the event of a natural disaster, robbery, arson, or some other unforeseen event. You can even insure expensive pieces of equipment to receive compensation if integral machinery suddenly stops working and needs to be repaired or replaced.

Commercial auto insurance is also a must. Remember, you depend on your company vehicles to get you to and from jobs. You cannot afford to not insure your vehicles properly. Look up local minimum auto insurance laws for your state and follow them.

Health insurance is probably the most expensive, and the most controversial, type of insurance on the market. Many HVAC contractors look at it as an unnecessary expense, something to be avoided at all costs. However, employees consider it an integral part of any job package, as evidenced by the fact that it is the most requested benefit among employees today. You do not necessarily *need* to offer it, but failing to do so could make it hard for you to attract the right workers to your company.

The forms of insurance outlined above can account for up to 7–8% of sales for the average contractor with annual sales of $500K to $1.5M.

Facility Overhead

The last step in the stairway of total overhead is having a place of business, along with the expenses associated to its upkeep. There is nothing flashy about these expenses. They do not drive business or offer any benefits to employees, but they are necessary because your business needs to have a physical address. Customers expect to see a company office that isn't based out of your home. This means that you need to spend money to buy, rent, or lease a spot. You need to pay insurance for your place, as well as paying utility fees. You also have to furnish your business office with essentials such as desks, chairs, and shelves. These expenses usually run up to 2% of sales.

The good news is that having a physical address can provide the company's owner with a variety of tax benefits. Industry data shows that discretionary earnings for the owners of HVAC companies with sales revenues from $500K to $2Mn a year range from 14% to 16%. Given that the average HVAC company owner's pay amounts to about 8% of sales revenue, the other 6% to 8% of earnings comes from other benefits such as vehicle use, company gas cards, or rental income if you own the building and rent it out to your business. All of these also provide some sort of tax relief.

Knowing Overhead Leads to Higher Profits

Understanding where overhead expenses come from and learning how to manage and even minimize these expenses when possible will enable you to sell jobs at a profit each time you are asked for a job quote. According to Bizminer, a financial benchmarking service, the average overhead for a contractor with an annual sales volume of $500K to $1.5M is between 37% and 40%. Most contractors in this category think their overhead runs between 20–25%, which is why so many of these contractors lose money on their jobs instead of turning a profit. On the other hand, contractors who know their overhead percentage and the market margin cap price average net profits between 10–15%; some even earn net profits as high as 22% on jobs sold. Remember, this is *net* profit we're talking about, not gross profit. This may sound impossible, given that the industry profit average currently hovers between 3% and 5%; however, when you realize that this figure not only counts the owners who make between 10% and 22% profit on every single job but also the owners who *lose* money on almost every single job, it is easy to understand why some company owners earn so much money in a field considered to be so difficult to turn a profit in.

Keep in mind that the numbers we use in the book for margin caps and overhead come from two sources; actual numbers may vary from region to region. The first source is our own consulting experience working with hundreds of contractors from the Carolinas, Georgia, Kentucky, and Indiana. The other source is a data-mining site called Bizminer. For this book, we focused on the Kentucky and Indiana markets, but Bizminer provides us with national numbers as well.

Another point regarding overhead should be mentioned here. In Chapter 2, I mentioned that there are different types of businesses represented in your own company. You may do residential system replacement work, service work, and perhaps residential new construction. I told you each of these segments had different margin caps affecting pricing in that segment.

Well, the same holds true for overhead. Each segment has its own unique overhead. This concept will be discussed more thoroughly in Chapter 4, but you need to be aware of it in this discussion. When I say the average overhead for small to mid-sized companies ranges between 37% and 40%, I am referring to the company's total blended overhead between all the different segments or departments. When we get deeper into pricing mechanics, you will see that you need to use the business segment or department's overhead that you are pricing work for.

Overhead: A Deeper Dive into Cost Reduction

There are almost always ways to reduce your overhead if you know where to look. Following are some recommendations that could help you cut costs without cutting back on the value of the goods and services you offer.

Operational Overhead

Unfortunately, you cannot dramatically reduce operational overhead expenses. You need financial and legal help to set up a business. There is simply no way around it. Without good legal and accounting help, you could wind up in trouble because there are many local, state, and even federal laws that could impact your business if you aren't aware of them. But that does not mean you cannot shop these services out. Online services like LegalZoom and Rocket Lawyer offer lower cost alternatives for incorporating, property transactions, and the like.

You also need a good computer to keep track of business income and expenses. Phone answering and secretarial services like Regus and Virtual Assistant are companies that can provide the tasks of an inside receptionist or secretary for a fraction of the cost. If you use or accept credit cards, shop the rates the bank or financial institution is charging you so you don't pay any more in fees that you must. As noted above, your operational expenses should make up no more than 3% of your total overhead costs.

Promotional Overhead

Many HVAC contractors balk at the idea of dramatically increasing their promotional spending. A common complaint that I have heard from numerous contractors is that spending on advertising does not work. Such contractors feel that all their work comes from referrals, so they do not spend money on promoting their business to new customers. However, there's a flaw in this line of thinking. I agree that in many cases, HVAC companies are not generating much, if any, business from advertising—but that doesn't mean that advertising does not work. In most cases, it simply means that the company is either not using the right method of advertising or is advertising to the wrong people.

The first step in starting a successful advertising campaign is deciding who to target. Many HVAC companies try to target all homeowners, home renters, and business owners who happen to live in their local area using ads offering equipment discounts and pre-season heating and AC tune-ups. The problem with these ads is that most of the individuals listening to them don't really understand the value of a biannual tune-up or simply don't care if their HVAC equipment is tuned up on a regular basis. Furthermore, many potential business and residential customers don't care about replacing an HVAC unit until that unit is past the point of repair. That's why handing out flyers, buying advertising space on TV, and creating direct mail campaigns usually results in net losses.

The key to a successful advertising campaign is to reach the people who want your goods and services. The best way to do this is to set up a top-quality website and actively promote your company online. You should also sign up for local directories such as Yelp and Angie's List. Claiming your business on either of these websites costs little to nothing, and these sites are often the first places that people visit when they are looking for HVAC services. If your business is focused on installing new HVAC units, get to know the custom home-building companies in your local area. You can do this both by paying in-person visits to local company owners and reaching out to them on LinkedIn. Once again, these forms of advertising are free but have the potential to generate a great deal of revenue.

I am not opposed to including top-of-mind awareness advertising as part of your marketing strategy but the messaging needs to focus on breakdown service. Your marketing needs to shout from the mountaintop that your business is the right choice for when people need service. Service breeds replacement and can create wealth faster than any other revenue stream in your business. That is why there is an entire chapter dedicated to the service gold mine later in the book.

As we touched on earlier, you probably need to spend more money on promotion than you currently do. However, if you own one of the rare HVAC companies that is spending more than 5% on advertising, it's time to trim your promotional budget. Take a close look at where your advertising dollars are going and estimate how much work you're currently getting from each form of advertising you pay for. Expensive billboard ads, television ads, and mail campaigns should be the first to go. If you feel that no one will know that you exist unless everyone in town sees your advertisements, hire someone to give flyers out for a small fee. I don't believe this is an effective form of advertising, but at least it's cheap and it can help people become familiar with you if you have just set up your company and are unknown in the local area.

I highly recommend that you focus on the free forms of advertising touched on in the overhead expenses section. The internet offers many ways to promote your business effectively and at low cost. A limited pay per click (PPC) campaign can also help if you set effective parameters so you aren't randomly spending money on keeping up with a competitor's campaign. Alternatively, you may want to forgo PPC ads altogether and look for ways to advertise on popular local websites. A static ad on a local site could do your business more good than a Google advertisement that most customers are unlikely to take a second look at.

The idea here is to reduce promotional costs by being more efficient with your advertising dollars—not to reduce promotional costs by refusing to spend on any form of advertising. Thanks to the internet and social media like Facebook and Twitter, there are many free forms of marketing available today. Now, remember: they may be free in terms of money, but they are not free in amounts of time. For this type of marketing to work, you must invest the time to make posts to your Facebook account and Twitter accounts. Social media works especially well as the millennial generation takes over the workforce; they'll account for 50% of all consumer spending by 2020. With that in mind, I recommend you become serious about your social media marketing.

Vehicle Overhead

Can you save money on vehicle overhead costs? Yes—if you choose the right vehicles in the first place.

Unfortunately, some business owners act like the 30% of customers we talked about earlier. They put cost ahead of potential savings and purchase the vehicles that look like the best deal without considering long-term costs. The fact is, some vehicles get far better fuel economy than others. Some brand names and models are better made than others and are therefore less likely to break down and need expensive repairs. What's more, some types of vehicles are even cheaper to insure than others. Do your homework so you don't wind up pouring money down the drain when you buy vehicles for your company. If you have inefficient vehicles, consider selling them and getting new ones. Doing so costs money but will also save you money in the long run.

Another important point I would like to make is that you need to have your vehicles serviced by a reliable, honest mechanic. Many contractors who are desperate to cut expenses rely on mechanics who are willing to do quick, subpar maintenance and repair jobs. In some cases, you'll need to switch mechanics; in other instances, you'll need to make it clear to the mechanic that quality service is more important to you than a cheap price tag and getting the vehicle back quickly.

With proper maintenance and care, your vehicles should last a long time even if you use them frequently. Improper care and maintenance will result in having to replace vehicles far sooner than would have been necessary otherwise. There are several fleet maintenance companies like ARI that can put your fleet on a regular maintenance program aimed at lowering your long-term maintenance costs.

Insurance Overhead

I am often asked to name the most draining overhead cost for an HVAC business. There is no doubt that, outside of payroll costs, this is insurance. In my opinion, workers' compensation and health insurance pose the greatest threat to both existing businesses that have survived their first four years in business and new startups.

This is exacerbated by the fact that insurance companies often consider new HVAC contracting companies to be a risk because they have no safety record. Expect to be charged a pretty penny for your first workers' compensation insurance policy. However, you can also expect the cost of your policy to decrease as the years go by as long as you maintain a high standard of safety and don't present compensation claims for the first few years.

The cost of a workers' compensation policy varies wildly depending on which state you live in. In Texas, you can expect to spend 75 cents on workers' compensation for every $100 you make. That's not bad; however, it is the bare minimum. In Alaska, a workers' compensation policy will take nearly $2.75 out of every $100 you make.

Health insurance is even more costly. As several reports note, the average employer-sponsored health insurance premium comes to nearly $6,500 per individual and just over $18,000 a year for families. This can add up to a lot of money if you have many employees.

Are there ways to bring these costs down? The good news is that yes, you can lower these expenses. However, first I want to tell you how *not* to lower these costs. We'll cover the details later when talking about the value of your workers to your business, but I want to strongly caution you here that cutting compensation to your staff is rarely a good idea. Your business needs experienced, certified HVAC workers to remain successful and you can only attract these workers if you are willing to offer generous benefits.

One way you can control payroll costs is to only hire the workers you need in the first place. Many HVAC companies I have worked with have one office worker for every one to two field technicians. This ratio is out of proportion. You don't need that many office workers; in fact, you should have no more than one office worker for every three to four technicians and/or installers. You may be able to reduce this expense even further by outsourcing the office work to part-time or flex employees who do not require the same benefits as your full-time office or field employees.

You may also want to take a close look at your own wages. As the business owner, you should be making a reasonable amount of money. However, the definition of "reasonable" is up for discussion. If you are starting a new business, don't expect to earn a high salary right from the start. Start with a living wage and then increase it as your jobs increase and profits grow. You may also need to reduce your salary from time to time if a temporary shortage of jobs occurs or if some unforeseen event brings about additional expenses.

Facility Overhead

There are several ways to minimize facility expenses without moving to a less desirable or run-down part of town. Start by taking a close look at your utility bills. Chances are you can eliminate waste there, lowering your facility expenses.

Next, look at the premises you rent or lease. Do you need as much storage space as you think you do? Perhaps you need to purchase a lot of equipment, parts, and supplies in advance. If that's the case, then you really do need lots of storage space for these items and this storage space must be safe, dry, and properly ventilated. However, not all HVAC companies need so much storage space. If most of your jobs involve servicing or repair work with seasonal equipment replacement jobs squeezed in between, you do not need to store lots of parts in advance. Order them as needed instead of in bulk. You may pay a bit more per unit, but you will save money in the long run by not paying rent for unused space. What's more, you will also avoid the problem of getting stuck with stock that you can't sell. Additionally, carrying less inventory will reduce your annual property taxes.

Finally, look at your office space. You probably don't need a large office if you don't have a lot of staff. Cut back on office space by renting a small yet good-looking office and you can save yourself a fair bit of money every year.

Chapter Summary

After this chapter, you should have a clearer picture of overhead items and the purpose they serve. If your expenses do not support a revenue-producing function, then they may be considered waste—and waste should be removed to keep your overhead as low as possible. The six Stairway to Total Overhead categories can be used as benchmarking figures to compare with your own financial data. Recognize that it takes discipline to keep reviewing overhead to find ways to reduce it and control it. Overhead is our arch enemy—one that can prevent us from being competitive or simply take away profits.

Chapter 4
Markups, Margin, Pricing, and Market Caps

Chapter Overview

In this chapter, you will learn:

- Common mistakes using markup multipliers
- Why the divisor method is more accurate
- Market cap ranges by business segments
- Custom homes vs. tract homes: Joe's story
- How to price for a more profitable service business
- Parts markups: Yes, I said it.

We have already talked about the difference between markups and margins. They are not the same—and considering them to be the same can result in losing tens of thousands of dollars in profit every single year.

As you have probably noticed, I am not a fan of the multiplier pricing method. While it *is* possible to use this method correctly if you know what to multiply and how much to multiply it by, it is usually used inaccurately. Therefore, I recommend that you discard it altogether. The single factor pricing method is the best method for pricing both new and existing HVAC equipment installation jobs unless the labor needed to get the job done will cost more than the equipment and materials used on the job.

The following table represents the results when the multiplier method is used. I have selected the most common markups used by contractors to figure a sales price for a consumer when pricing a replacement equipment job or marking up service parts. The most common mistake here is when the contractor believes the markup is the margin she will get from the job after the job costs are subtracted from the sales price. From the table, you can clearly see this is not true. Markup does not equal margin.

Common Mark-Ups and Their Gross Margin Percentages

Mark-up	Job Cost	Expected GM Margin %	Actual GM Margin %	Proof Table
25%	$2,500	25%	20%	[1.25x$2500=$3125] [$3125-$2500=$625] [$625/$3125=.2 or 20%]
30%	$2,500	30%	23%	[1.30x$2500=$3250] [$3250-$2500=$750] [$750/$3250=.23 or 23%]
35%	$2,500	35%	26%	[1.35x$3500=$3375] [$3375-$2500=$875] [$875/$3375=.26 or 26%]
40%	$2,500	40%	29%	[1.40x$2500=$3500] [$3500-$2500=$1000] [$1000/$3500=.29 or 29%]
45%	$2,500	45%	31%	[1.45x$2500=$3625] [$3625-$2500=$1125] [$1125/$3625=.31 or 31%]
50%	$2,500	50%	33%	[1.50x$2500=$3750] [$3750-$2500=$1250] [$1250/$3750=.33 or 33%]

The Divisor Method and Why Division Is More Accurate

The divisor method formula I recommend you use is the single-factor pricing method we discussed earlier. If you remember, the formula is [Direct Job Cost / 100% - Desired Gross Profit %].

However, when you look at the formula, you will notice that the overhead percentage is missing from the calculation. You need to know the overhead percentage to determine what the desired gross profit percentage must be as the denominator in the single-factor pricing formula. To figure what the denominator must be, you must first take the overhead percentage and add it to the net profit percentage you want to make on the job.

Imagine, for example, my company overhead is 35% and I want to make a 10% net profit on the job. Knowing these two pieces of information, I can figure out what my desired gross margin needs to be by simply adding them together. A 35% overhead plus a 10% net profit goal equals a 45% desired gross margin—that's what's needed to give me a 10% net profit on the job after my 35% overhead is taken out. Remember, we must subtract this 45% from 100% to arrive at the divisor or denominator to be used, which in this case would be 55%.

The math looks like this. Let's say my job cost for labor, all materials, and any extras comes to $4,300. My single-factor pricing formula would look like this: [$4,300 / 0.55] for a sales price of $7,818.

You can prove the formula is right by taking the $7,818 sales price I mentioned above and subtracting the $4,300 direct job cost. You then have a gross margin of $3,518, or 45% of the total cost. It sounds like you made a bundle of money—but now you need to subtract your

35% overhead. Remember, one of the most common mistakes contractors make is identifying gross profit as net profit. Gross profit is *not* net profit. Once you take out the overhead from the gross profit number ($3,518 - $2,736) you have a net profit of $782, or 10%. I got the $2,736 overhead number by multiplying my 35% overhead by the $7,818 sales price.

The single-factor method is the most assured way of accurately deriving a sales price that meets a company's gross profit and net profit goals when the overhead percentage is known.

Market Caps: How Much Should You Charge?

How much is too much?

This is a question that plagues HVAC contractors all over the country. While there are a few who charge exorbitantly high prices for their services, this is a very small minority. The overwhelming majority of HVAC contractors simply do not charge enough to stay in business or make a reasonable net profit.

Knowing how much the market is willing to pay can be difficult. There is no set number; the results vary depending on where you live and what services you provide. Even so, this in-depth look at each construction market can give you an idea of what you may be able to charge for your services.

New Residential Construction

It may surprise you to realize that new home construction companies, much like HVAC companies, usually do not have high net profit margins. Companies that concentrate on larger custom homes tend to have higher net profit margins than companies that focus on smaller tract houses, apartments, and condominiums. Even so, the net profit margin for residential new HVAC companies ranges from 2% to 4%.

The residential new construction industry is very cutthroat and competitive. As builders accept bids at any time in an effort to reduce costs, the new construction business is the easiest market for startup HVAC companies to enter the field because they do not need to make an investment in advertising to get started. Now, I'm not saying that you need to lower your bid and lose money on a new home construction project simply because the builder is trying to buy the job down. The point I want you to bear in mind is that new residential construction is a market where your customers are actively trying to get the best possible price. If you are seeking work in this market, expect them to look for ways to cut costs. These clients will be getting bids from different contractors, and in many cases, they will go with the lowest bid offered unless given a reason to do otherwise.

So how much can you expect to earn on a new residential construction job? In markets we have consulted in, a typical three-bedroom, 1,200- to 1,500-square-foot tract home will go for $7,500 at roughly a 35% gross margin. Bear in mind there will be lower bids, but if you want to stay profitable in the new construction tract home business, you are in the $7,000 to $8,000 ballpark.

Custom Homes: Why They Make a Difference

Though tract builders like Pulte Homes, Arbor Homes, Dominion Homes, and Centex Homes offer a lot of work because of the volume of homes they build on an annual basis, their demands on contractors are high. To succeed in the tract business with its low profit margins, a contractor must have a business model designed to handle high work volumes. What this looks like can vary somewhat, but in most cases, it is done by pairing an experienced lead mechanic or journeyman with lower wage helpers who do most of the grunt work.

The real investment in this market is the number of people and trucks required to keep up with the volume of work. More people mean higher benefit costs, primarily in terms of workers' compensation, auto, and property casualty insurance. The need for trucks can eat up working capital if paid for in cash or, if financed, can eat up cash flow and increase debt service as you pay off the loans. With this model, the company must do high volumes of work and be paid in a timely manner to maintain cash flow and be reasonably profitable. Therefore, most small contractors cannot survive in the tract business—I don't recommend they participate in this market.

Another thing to consider with the large-volume tract business is that builders are demanding of your time, and if you cannot deliver, they will find someone else. This demand creates a huge conflict for HVAC contractors, as builders are busiest in the summer when the weather is conducive for building—yet this is the same time the very lucrative existing home system replacement business is at its peak. This scenario forces HVAC contractors to decide on which market to serve. Because of resource limitations, high-volume tract HVAC contractors are often forced to walk away from the replacement business.

There is an answer for HVAC contractors who want to focus primarily on the replacement business yet feel the need to have some form of business that picks up the slack when the replacement business cools off. The answer is custom homes.

The custom home market offers several advantages to the smaller HVAC contractor. The biggest advantage is a custom home delivers a higher profit margin to HVAC contractors per house than tract homes do. Custom homes make it easier to share resources between that business and replacement sales opportunities during the peak summer season. Custom homes usually move at a slower pace than tract homes, so a contractor can pull a crew off a custom home to do a quick HVAC system changeout in an existing home. This flexibility to mix two higher profit margin segments of the HVAC business allows contractors to be more profitable.

That profitability comes from two sources. First, the market cap for these two segments is much higher than for the tract home segment. Gross margins for the custom home market range from 38% to 40%, while the replacement market cap can range between 45% to 50%. Second, a contractor does not need to invest huge sums of money into trucks and employees to participate in these market segments at the same time. These cost reductions and lower associated insurance costs also help to improve profitability.

Choosing between tract housing, custom homes, and existing home replacements is a personal choice. This discussion is not meant to persuade you about how to build your business model, but merely to point out the investments, risks, requirements, and rewards of each market segment.

Even so, this advice does not apply to all companies. Perhaps you have just started an HVAC business and do not have the cash flow to carry slow-paying contractors. Maybe you have a small crew and cannot afford to hire the contractors you need to get a large residential HVAC job done on time. If these scenarios describe your company, you may want to reconsider getting involved in the residential construction market in the first place. If you are serious about breaking into this market, though, make sure that you are turning a profit on your projects by accurately estimating how much each job will cost. Consider the cost of materials, how much you would need to pay your technicians, and the transportation costs involved.

It is also important to choose your residential construction customers with care. Remember what I outlined in Chapter 1: it's not really a jungle out there. You *don't* need to accept every single project that comes your way. Many HVAC companies lose money when working with residential construction customers because they look at the large price tag on the project and feel certain that they will turn a profit on such an expensive job. Unfortunately, it often does not work out that way. You may be earning a very large gross profit on the job while also incurring large expenses that aren't covered by your earnings.

If you are interested in breaking into the residential home construction market, get to know the companies in this market very well. You need to know where the construction will take place. Don't assume it will all be in town, close to your business office. Extended travel time can raise both fuel overhead and labor overhead, as you simply cannot refuse to pay your workers for the time they spend traveling to the construction site. Finally, you need to know what will happen if there are on-site delays. Who pays when your workers spend hours sitting and waiting for one aspect of construction to be completed so they can start work? It shouldn't be you.

As you can see, this market is complex. If you aren't sure what you are getting into, don't bother with it. Other markets have the potential to not only be more profitable but also simpler to manage.

Crunching the Numbers: How Much Net Profit Should You Expect?

If you keep your company overhead no higher than 32%, you should be able to turn a profit in the new home construction business. Knowing that the market cap for tract housing is roughly 35% while custom homes range from 38% to 40%, you can use the divisor formula outlined earlier in this chapter to determine how much you can charge for each job to turn a profit. Using these parameters, your tract housing will generate a 2–3% net profit whereas the custom work will generate 6–8%.

It is possible to make good money in residential construction if you know the financial principles outlined in this book. If you don't believe me, consider the case of Northpoint Heating and Cooling. Joe loved new construction and existing home replacements, which always baffled me. I often asked him, "Do you really want to do these houses when you make so much more money in the replacement business?"

The answer was yes. Joe had a passion for houses, but he also had a passion for the 12 pillars of success. He especially had a good grasp of the different types of overhead between his new construction department and his retail replacement business, allowing him to submit accurate price quotes for each business segment. He was smart and knew how to make money in new construction.

Joe focused on custom homes, which was a good fit for his area. At the time, the Columbia, South Carolina, market was growing. With a lot of money coming in and a lot of new transplants, there were a lot of custom homes being built and he became the custom home king.

What's more, he also learned what it takes to become a great retailer. Joe made regular TV appearances and became a well-known persona in the HVAC industry in his area. He had a solid reputation in the custom home new construction business and developed a high-visibility profile in the lucrative retail replacement and service business. By applying the single-factor pricing method and understanding the market caps for each of these segments in the Columbia area, Joe has come to own one of the most profitable HVAC companies in the area with annual net profits well north of 15%, no pun intended.

Joe's success has enabled him to enjoy the fruits of his hard work. He had a love for Lake Murray, and through his profits, he built a beautiful lake home that he and his family enjoyed as the kids grew up. Later, Joe's solid reputation in the custom home market provided an opportunity for him to follow one of his custom home builders to Beaufort, South Carolina, where he opened a satellite operation that his oldest son Ryan now runs. Joe's success there allowed him to build another beautiful home on the water.

Why was Joe able to do all this stuff? Because he mastered the principles of success that I'm telling you about in this book. He learned all these concepts that we're talking about and then applied them to every single aspect of his business. The great thing for Joe is that he has lived the dream of every HVAC contractor who decides he is going into business for himself. He started by turning the wrench, then added people, followed by building and managing a very successful business. Today, Joe is financially set—proof that by knowing and applying the 12 pillars of financial success, the American dream really can come true.

Scott Ritchey with Gary Kerns

Residential Service

Residential service, though often thought of as evil by some, turns out to be the best driver for higher net profits and the catalyst for a healthy and very profitable retail existing home replacement business.

Homeowners in every single city across the United States rely on air conditioning in the summer and heating in the winter. Yet not all homeowners understand the importance of periodic maintenance, which is a shame, as regular maintenance could keep an HVAC system working longer and more efficiently than would otherwise be possible. Even so, it would be hard to find a homeowner who isn't eager to have a unit repaired ASAP if it stops working.

The good thing about residential service is that it creates more work for you. By developing a strong reputation for providing quality service at a fair price when repairing broken-down HVAC equipment, you ensure homeowners are happy with your work. They will then refer other potential customers to your business.

When I ask contractors where their work comes from, the overwhelming response is "referrals." This is true, but referrals from customers don't fly around like banners behind airplanes. They usually happen when a friend or relative needs your service. I mentioned earlier that consumers do not have the same relationship with HVAC contractors as they do with their doctor or dentist. Because breakdowns are infrequent, you will soon be out of sight, out of mind if you are not in constant contact with your referral network. Therefore, it is imperative for long-term success to develop programs that keep your company name in front of them.

There are many low-cost ways to do this. You might gather customer birthdates to send birthday wishes, either with a paper birthday card that has a personalized note inside or via email. Another form of contact might be service anniversary cards that include discounts on other services you offer that are sent on the anniversary of the last service call made to the home. Annual fall and spring preseason service reminders are another way to reach out to your customer base, keeping the referral network alive and well.

No matter how you choose to do stay in touch, be like Nike and "Just do it."

MAKE MORE MONEY: 12 PROFIT PILLARS FOR HVAC CONTRACTOR SUCCESS

Service Agreements: A Referral Treasure Trove

All the above ideas are ways to keep your referral network strong, but none of them are as lucrative as a robust service agreement program. Residential service agreements can last for one or more years; how much they cover will depend on the services you offer and what your customers are willing to pay.

Some customers only want a basic maintenance agreement that provides equipment inspections or minor service procedures like outdoor coil cleaning and thermostat calibration. The main benefit of this type of agreement is that it keeps the company name in front of the customer twice a year and keeps service techs working in slower times of the year.

A full service agreement, on the other hand, can be very profitable, as such an agreement not only includes replacing broken parts but even putting in a brand-new HVAC system when necessary. If you are going to offer such an agreement, you must know the age of the equipment, repair history, and risk that such an agreement brings. I only recommend offering these types of agreements on new installations that you performed; I also recommend that you limit offering a brand-new unit replacement to the first eight years of the unit's life, and only if the repair cost would be greater than $2,500. For additional protection, most equipment manufacturers now offer 10-year parts warranties if you fill out the proper registration. Complete the registration and the manufacturer bears the cost of the parts under the full service agreement for the first 10 years. Most full service agreements range in price from $350 to $500 annually.

The main thing that stands in the way of getting these agreements is the fact that many homeowners do not like to pay for services that they don't understand the importance of. However, this is not an obstacle that has to hold you back. Most residential customers place more importance on comfort and long-term energy savings than they do on price. If you are willing to deliver quality service in a timely manner, your customers will be willing to pay well for it. Even though the average service ticket is between $250–$300, the gross margin percentage on this work ranges between 60–65%. The service business is very lucrative and is the breeding ground for system replacements which is covered in the Service Tech Goldmine chapter of this book.

Scott Ritchey with Gary Kerns

Common Repairs and Their Associated Prices

Improvenet recently came up with a handy list of how much HVAC companies typically charge to repair certain types of damage:

- R-22 Refrigerant Leak Repair: $200–$1,500, depending on evacuation and recharge
- AC Refrigerant Recharge: $250–$750, depending on type of refrigerant and the # of pounds
- Control Circuit Boards: $350–$800, depending on brand and PSC vs. ECM motor
- Fuses, Circuit Breakers, or Relays: $75–$290
- New Thermostat: $115–$750, depending on features like WiFi, touchscreen, etc.
- Outdoor Condensing Unit Fan Motor: $450–$650 for universal; add another $400 for OEM.
- Condensate Drain Line Flush: $75–$250
- Compressor Replacement: $1,750–$2,500, depending on refrigerant type and reason for failure
- Replacement Condensing Unit Coil: $1,900–$2,900, depending on coil size and refrigerant
- Metal or Plastic Drain Pans: $250–$575
- Condensate Pump Replacement: $240–$450

As you can see, there is a wide price range for each repair job. You want to set your prices as close to the market cap as possible while still giving customers a reason to purchase your services. However, don't just take this list and base your prices on it without thinking about your business, how it operates, what you are paying your workers, what type of HVAC unit you are working with, and all other related factors. You also need to consider where you live, which is a point outlined in detail below. For instance, people in California are probably far more willing to pay a high price for HVAC repair than someone living in a state with a much lower cost of living.

Geographic Location

If you aren't sure what the market cap is in your area, do a bit of research. A number of helpful websites can make it easy to see how much companies are charging in your area.

Home Advisor has a handy website (https://www.homeadvisor.com/cost/heating-and-cooling/repair-an-ac-unit/) that makes it very easy to see the price range for HVAC repair in any given city. Type in your ZIP code and you will not only see the average (which you should never base your estimates on) but also the lowest and highest prices offered. You can be sure that the successful companies are the ones with the higher price range, whereas the unsuccessful companies are the ones pushing prices lower.

The website (http://www.improvenet.com/r/costs-and-prices/air-conditioning- repair-cost) also has a handy cost estimator. This one is a bit more specific because it asks you to specify what type of repair job you want to know more about. You put in your ZIP code and then click on the appropriate repair job. The program then shows the market cap in your local area.

I would advise you to bookmark these two websites and check them at least a few times a year. But remember, a number of factors can affect the market cap in your area. Supply and demand will always have a bearing on market prices; if a number of small HVAC contractors offering cheap work move into your area, then average prices will go down. A boom in home construction could boost prices, as this line of work creates multiple HVAC jobs. You can either move into job openings created when other HVAC companies focus on home construction or move into the home construction market yourself.

The overall economic condition in your area will also affect prices. If you notice that people who had been happy to pay your prices in the past are no longer quite as willing to do so, take a moment to step back and have a look at the financial news. You may need to lower your overhead so you can lower prices without cutting into your profits. It may even be time to consider branching out to another city with better job prospects.

Never coast on your laurels when working in HVAC. Mind you, I'm not talking about overworking to the point of exhaustion. I know that nearly all HVAC business owners are hard workers. They work from morning to night on weekdays and, in many cases, they even work

Scott Ritchey with Gary Kerns

eekends. What I mean is that you cannot afford to rket cap once and then leave it at that. The figure d you may very well need to change with it.

back to the point we talked about earlier: you need to step back from time to time to think about how your business is doing. Look for ways to save money without reducing services; consider which types of work you want to do and what types of jobs you would rather avoid. Evaluate employee productivity, office needs, your vehicular overhead, and all the other factors affecting your company's productivity.

Pricing for a Profitable Service Business

Typical overhead for a service department ranges between 40–45%, whereas a replacement department's overhead will range from 32–35%. Therefore, pricing principals and mechanics differ greatly between the two. The first difference to know is that the single-factor pricing method was designed for installation-type jobs. Using this method ensures that retail salespeople get the proper gross margin they are looking for after they figure out the variety of costs that are needed to do an installation job. In service transactions, there are only two costs associated with the transaction: the technicians labor and the cost of the repair part.

The second difference is that the smaller the part, the higher the gross margin that can be made by the contractor because of the lower parts cost that still allows you to offer a reasonable price to the consumer. As a result, the best method for pricing service parts is to group parts cost into specific cost ranges and use a multiplier markup to ensure that the gross margins for service parts average 55–70%. This may sound high to you, but these are the required margins necessary to return a 15–20% net profit on a service department that has an average overhead between 40–45%.

The reason I prefer the multiplier method for service is that it's easier for service technicians to work off a pre-determined parts markup schedule for both pricing accuracy and speed and efficiency when closing out a repair ticket invoice. A common complaint by consumers is that they feel the technician is not performing any service while they are filling out the ticket, so they should not be charged for that time. Having a preset labor rate and parts markup schedule allows this process to be performed much more quickly, in a way that's acceptable to the consumer.

When it comes to service repair work pricing for consumers, many companies have turned to flat rate pricing programs. These systems usually charge the consumer a diagnostic fee ranging from $50 to $80 for the technician to determine what's wrong with the HVAC system, then offer to repair it for a flat fee that includes the labor and parts costs. This offers no hidden fees or cost overrun charges if the technician takes longer than normal to repair the system.

Though we do not endorse a particular "flat rate" system on the market, we believe it is a sound business practice to use such a pricing methodology for your service work. If you do not want to use a flat rate model for your business, then we suggest you follow these guidelines.

To set your labor rate, aim for a gross margin return of 65% for your service labor. To do this, take your highest paid technician and multiply his or her hourly rate by 3. For example, if your highest technician's hourly wage is $25, you would multiply this by 3 and come up with a service rate of $75 per hour. It is very important that you keep in mind that this rate *only* covers his or her hourly wage and the labor burden benefits that are associated with that technician. You can always figure a labor burden of 30% to 35% per employee for insurance benefits, payroll tax matching, paid vacations, and retirement benefits like 401(k) matches if offered.

We strongly suggest you charge some sort of service fee to cover the costs of the truck used to transport the service technician and the associated costs of insurance, fuel, and maintenance repairs. You can call it a trip charge, service call charge, or diagnostic fee. The critical message is simply that you must charge *something*. Usually $40 to $60 covers the cost and is acceptable to the consumer. The bottom line is, if you are not going to use a flat rate model, then your service rate needs to be at least $100 for the first hour to make the necessary margins to be profitable.

When it comes to marking up parts, use the following guidelines to achieve the correct gross margins needed to operate a profitable service department.

Parts Cost and Markup Schedule

$0.00 to $4.99: multiply by 4

$5.00 to $9.99: multiply by 3.5

Scott Ritchey with Gary Kerns

$49.99: multiply by 3

$99.99: multiply by 2.75

$).00: multiply by 2

Chapter Summary

We covered a lot of detailed information in this chapter that is integral to the success of your HVAC company. Let's go over it one last time and look at the most important things you need to remember to consistently turn a profit on your jobs.

Markup and margin aren't the same. Never forget that! Work with percentages and the divisor method to accurately calculate installation job prices. Also keep your current market cap in mind. Don't settle for the bare minimum; charge what your market is willing to pay.

Knowing each aspect of your business is also important. In some cases, this will be easy. If you focus on only one or two types of work, then you won't have a hard time calculating your overhead for each type. If, on the other hand, you run a multifaceted company that handles all sorts of construction projects, you will need to take the time to calculate your overhead for each one to ensure that every single job is profitable.

Knowing your overhead for each type of construction project can also help you make wise overall business decisions. If you find that one line of work is not very profitable and/or it takes a lot of resources to generate relatively low profits, don't be afraid to drop it. My coaching partner Gary did that when he looked at the new residential and light commercial construction side of his business and decided to totally reinvent himself. It was a huge undertaking because he basically sacrificed $600,000 to $700,000 of his business by turning his back on these type of construction jobs, saying, "I'm not going to do that business anymore."

Gary understood the financial pillars of success before making this move. It was not a decision he came to lightly. He knew both the risks and the potential profits of making this move. The results were extremely positive. He later told me that his business grew by 20%-- but a closer look at the numbers reveals that it grew far more than that. Because he made up the $600,000 he lost working in the residential and light commercial construction sectors with new existing home residential replacement and service work and boosted his business by an additional 20%, the total growth percentage for his business was an astounding 120%.

Scott Ritchey with Gary Kerns

Chapter 5
Your Replacement Business Opportunity

Chapter Overview

In this chapter, you will learn:
- Residential service breeds replacements
- How to size up your replacement opportunity
- How to develop a replacement growth strategy
- Consumer financing as a closing tool
- Commercial service breeds replacements

Residential HVAC Replacement

Having a strong residential service business is very helpful when growing a strong residential replacement business. If there is one concept in this book that you should make a core focus in your business, it's that *service breeds replacement*. In fact, companies that understand this will have a ratio around 25% to 30% service revenue to replacement revenue. To calculate this, just divide your service revenues by your replacement revenues and you will find your service sales to replacement sales ratio (Service Sales / Replacement Sales = Ratio %). However, it is important to remember that although the service business and replacement business interact very well together, they are completely different in the operational cost it takes to support them.

There is one similarity between the residential HVAC repair market and the residential HVAC replacement market: both markets have the potential to bring about more work. If you do a good job replacing an HVAC system for a customer and the customer is happy with your work, you can expect to get called if something happens to the unit. Making warranty repair calls a priority sends a message to your customer that you stand by your work and they chose the right contractor to form a relationship with.

Some homeowners don't bother maintaining their unit or calling for biannual maintenance. Others don't use the unit wisely and so wind up putting a lot of wear and tear on it. This lack of concern by the homeowner is where service can breed replacements, making service and replacement inseparable.

How Big Is Your Replacement Market?

Crunching the Numbers

I often ask a contractor who wants our consulting services, "Do you know what your replacement sales share of market is?" Truth be told, in 33 years, I have never had a contractor give me an answer that was backed up by any benchmark or formula to prove some sort of accuracy. The reason I ask is not really to get an accurate answer but to see if the contractor has any idea of his company's potential for replacement sales in the geographic market served.

If you want to grow your business, the first step is to understand how big your market is so you can capture the amount of business your company can handle and deserves. Waiting for the phone to ring because you have a shingle out front with a company name and a business telephone will only make you a member of the "47% failed by year 4" club or one of the 90% who fails by year 10. But if you know how large your market opportunity is and how the service and replacement business work together, you will achieve the market share your efforts allow you to earn.

There are several ways to find out how large the market you serve is. Equipment manufacturers provide industry data to their distributors showing how many units have been sold into that market. You can ask your distributor for this data—they'll usually share it. However, this data is not as granular as you might need to see what's really specific to your business. For example, a distributor in Indianapolis can tell you how many units have been sold in Indianapolis as a whole, but they cannot tell how many were in sold in the west side of the city if that is the only part of Indianapolis you serve. This keeps you from knowing your replacement unit share of market in your specific area.

Another way to look at potential market share instead of units sold involves examining the dollars spent on HVAC products in your marketplace. This is calculated by a formula called BPI, or Buying Power Index. This formula takes the average amount of dollars spent by a homeowner per year in the HVAC category and multiplies it by the population of the area. Your market share is then calculated by taking your annual sales revenue and dividing by the annual spent dollars in your area provided by the index. The drawback here is that this does not tell you how many replacement units are available for your company in your area. Not knowing this hinders your ability to create a growth strategy to reach a specific number of units, nor does it help you figure your replacement unit share of market.

When we consult with companies, we use the industry average life of a condensing unit or furnace and the number of homes in the market served by the contractor to figure the replacement market potential and the contractor's replacement share of market. A quick disclaimer here: though not scientific, this method is close enough for contractors to figure out their replacement sales potential and develop a strategy to achieve it.

The first part of this method involves determining how many homes are in the area. To make a quick calculation, if the contractor is in a rural town or county, we take the population and divide that number by 4, because we figure four people to a home constitutes today's average family size. This comes from the American birthrate statistics of 1.9 children per family. Once we have the number of homes, we deduct the number of apartments. This data can be found in county census data, freely available on the internet. This data will also provide the number of single-family homes, so you do not have to use our calculation method if you don't want to.

As it turns out, though, this only works if you are serving a rural county and the small towns within it. If you are in a large metropolitan market and do not serve the entire area, then the census data will not work. There is a way, however, to figure the number of homes in your area if you are in a metropolitan area, but it requires some work. Believe it or not, your local post office can help you find this information. Each mail delivery person is assigned a carrier route number. The post office can tell you where those route numbers deliver and how many homes are in each area. Having this information will allow you to use the rest of the formula to figure out how big your replacement market is and how you can attack it.

Manufacturer data and contractor experience tells us the average life of an outdoor condensing unit is 14 years. Knowing this, we can estimate that an outdoor unit installed 14 years ago is likely to be replaced in the coming year. As such, an outdoor unit installed 13 years ago is due for replacement in two years and an outdoor unit installed 12 years ago is due in three years and so on. Following this logic, 1/14 of the units installed over the last 14 consecutive years are due for replacement. What this means is 1/14, or 7%, of the condensing units in your market served are going to be changed out this year.

If a market has 2,000 homes, then 7% of those, or 140, will need their outdoor unit replaced in the current year. To figure the share of market, simply take the number of outdoor units you changed out this year and divide that by the total number of units changed out in the market as a whole to get your outdoor unit share of market.

In this example, if I changed out 25 units, then my market share is 25/140, or a 17.8% share of market. Another way to look at this is that my competitors have an 81.2% share of market. Wow, what an upside I have to capture more market share! Knowing this, I can develop a strategy to increase my number of outdoor units sold and grow my market share.

Let's look at the gas furnace market next—the same principle applies. Industry data shows us the average life of a gas furnace is 20 years. This means that 1/20, or 5%, of the gas furnaces in your market will be replaced in the coming year. In the market example of 2,000 homes, this means 100 gas furnaces will be replaced in the coming year. If I replaced 17 gas furnaces, then I have a 17% share of the gas furnace market in my area (17/100).

Developing a Replacement Growth Strategy

There are two types of strategy to use when growing your replacement business. You can harvest your existing customer database, especially if your recordkeeping tracks the age of customer units that have been serviced by your company and the dates when your company installed new systems to calculate what units are 14 years old and older. Since these folks are already your customers, they should have no problem with you contacting them directly to see if they need any of your services.

There are several ways to do this. Many top companies have gone to outbound calling to promote specials on services designed to get technicians into homes where they can provide the service and check the condition of older units. In many cases, these visits uncover if the homeowner has been contemplating changing out their old unit.

It is important to note here that the technician is not going out with the intention of selling a new unit. The first problem with this is that most technicians do not like selling, nor do they want to be perceived as a sales person. Second, your business strategy should never involve pushing customers into making a purchase decision. They will make those decisions on their own. The point here is that it's important for you to be in your customer's top-of-mind awareness as they get closer to the time to replace their HVAC system. If you are just waiting for the call to come 14 years after you installed the unit, good luck. Remember that "out of sight, out of mind" thing again. If you do not have the customer under a service or maintenance agreement, chances are they will use someone else—especially if that someone else has been advertising to them in your absence.

I'll cover this next scenario in more detail in Chapter 9, but I want to touch on it here, too: quiet market share sale. This situation is why it is important to keep in touch with your existing customer base whose outdoor units are 14 years or older. While the service tech is in the home performing the service, a homeowner who is thinking about replacing his or her unit will ask that technician some questions that may sound like they are interested in buying a replacement unit—but this is merely a trap designed to get the technician to give the homeowner a ballpark price that they will use against you to shop for another price.

We teach technicians how to avoid this trap and turn the scenario into a quiet market share sale opportunity through a technique called the pendulum swing. Look, the homeowner only went on a fishing expedition because they really are experiencing some level of pain, whether it has been some nickel-and-dime repairs or because their utility bill has reached a breaking point. They want to know how much a new unit will cost them, but they won't want to buy right then unless they come to that conclusion on their own at that moment. This is why we call it quiet market share. If the technician executes the pendulum swing with precision, the homeowner will purchase a unit right then, and none of your competitors would ever be aware that unit was available to bid on. Hence, you stole one from the market, or quiet market share. We'll cover this process in greater detail later in the book, but for now it's enough to know that you need to stay in touch with those quiet market customers who may soon be looking for a new unit. In addition to an outbound call center for harvesting existing customers, you can use email blasts, direct mail offers, or service reminder postcards to do so.

The second strategy to grow your replacement business is to target prospective customers who do not currently do business with you. There are many forms of media to choose from but some are more effective than others, especially if you don't have a large advertising budget. If money is no object, then select several media options. TV and radio will be successful in getting your company name recognized if you have a lot of money to invest in them, but they are very weak at getting direct responses or calls to action if you advertise a promotion or sale on them. As I stated before, you are in a demand business. If customers do not need your service at that moment, they will not take you up on the offer.

There are better choices that are lower in cost and more successful at matching up consumer demand for your services. We are big fans of using internet marketing and mobile-ready websites. This is the most effective way to match up consumer demand with your company and its services. When faced with unexpected breakdowns of their heating and cooling systems, consumers reach for their phones and the power of Google to offer solutions to their problem. Therefore, we recommend that all of our clients have a well-designed website that uses a strong search engine optimization (SEO) company to deliver a high number of hits. This lets our contractors get the number of leads they need to meet their growth objectives. SEO fees range from firm to firm, but unless they can prove a significant number of hits over the average, you can expect the fee to run somewhere between $300 to $400 per month. These fees are for SEO only—this technique is designed to get you on the first page of a Google search, which usually lists 8 to 10 companies per page. Studies have shown that 92% of people using Google to do a search only look at the first page. Therefore, it is critical to get on that first page.

In addition to SEO management and internet marketing, another tactic is pay-per-click (PPC) advertising. PPC spots are paid advertisements that appear at the top of a results page in the shaded box. Usually there are three to four contractor website choices listed here to be clicked on. A good pay-per-click campaign starts at about $500 per month, but obviously the more you can spend, the better your results will usually be. Large contractors spend typically between $1,000 to $1,500 monthly on pay-per-click.

Internet marketing firms like Yodle that specialize in providing SEO management programs for companies suggest that a $600 per month SEO and pay-per-click 90-day program will generate 40–60 leads.

This may not sound like much, but consider these facts regarding leads and the potential sales that can result. The typical closing rate for the HVAC industry on replacement sales leads is 28%. Looking at the Yodle data, this means you'll close about 28% of the 40–60 leads. We already discussed that the average replacement sale is $6,500. Applying the math, the sales potential under this scenario would look like the following:

- 40 replacement equipment leads times 28% = 11 closed sales
- 11 closed sales times $6,500 average sale price = $71,500 revenue
- $71,500 sales revenue at 45% gross margin = $32,175 gross profit
- $32,175 gross profit – $1,800 program cost = $30,375 net gross profit

Please understand that this hypothetical only works if the keywords used by your pay-per-click ads and SEO focused solely on searches pertaining to HVAC equipment replacements. An example would be a search that said, "need new air conditioner" or "replace old air conditioner." A phrase like "air conditioner broke" is too generic—the person may just be looking to have their air conditioner fixed, not replaced.

The key here is that once you know how large your replacement market potential is, there are all kinds of tactics you can use in order to grow your share and sales volume in the replacement business. The residential replacement business is the most lucrative segment of a residential contractor's business. While it's true that service repair business offers higher gross margin percentages per transaction, the gross margin dollars generated per service transaction pale in comparison to the gross margins generated by a replacement sale transaction. The average service transaction of $250 with a 65% gross margin generates $162 gross margin dollars, whereas a replacement sale of $6,500 at a 45% gross margin generates $2,925 gross margin dollars.

As it turns out, out of the hundreds of contractors with whom I have consulted, those who had a firm grasp that service breeds replacement had the healthiest income statements and balance sheets I have observed.

A 10–15% net profit margin on an HVAC residential replacement job is very reasonable.

The typical national price range for replacing an old HVAC condensing unit and coil is between $3,500 and $6,600, depending on whether it is an air conditioner or a heat pump. Obviously, the tonnage and SEER rating of the unit play a factor in its cost as well.

Consumer Financing: The Most Overlooked Closing Tool

Consumer financing has been around for decades in the HVAC industry, yet only a small percentage of contractors use it as a closing tool on replacement system sales. Those who aren't are missing out. A 2010 Decision Analytics study found that 37% of homeowners walk away from a contractor's proposal because the contractor did not offer them financing. Think about that. Thirty-seven homeowners of the last 100 that you made HVAC replacement equipment proposals to walked away from you because you did not offer financing!

This is because the average savings account in America has roughly $500 to $800 in it. We must remember that more than 75% of HVAC replacement equipment sales were not planned for. The ability to finance an HVAC system purchase is a great relief to many homeowners who have just received the bad news that their system must be replaced. Homeowners who need retail financing usually do not have enough equity in their home to borrow on a home equity line of credit, or their credit score might be below the bank's risk rules for such loans. Retail financing companies like Synchronicity, Green Sky, and FTL all offer financing alternatives for this 37% of consumers.

Consumers who finance are a treasure trove for sophisticated contractors who offer Good-Better-Best option selling systems. Most of these retail financing programs have a minimum monthly payment of 2% of the amount financed; the loans are usually paid off in 84 months. Such low monthly payments allow the homeowner to make system upgrades for efficiency and indoor air quality very easily and affordably.

Take the average 14-SEER air conditioner and coil replacement at $3,500. The minimum payment on a 2% repayment plan is $70 per month. For an extra $1,000 total, the consumer could upgrade to a 16-SEER air conditioner, paying an additional $20 per month for a new total payment of $90 per month. The two-point increase in SEER rating will reduce the homeowner's cooling cost by roughly 20%. A homeowner with an average annual cooling cost of $800 would save $160, or $13 a month, basically allowing that homeowner to upgrade for only $7 per month.

Let's say the homeowner is interested in indoor air quality or having protection against carbon monoxide poisoning. We'll use the 14-SEER replacement for $3,500 and add an 11 MERV-rated air filter and a Kidde carbon monoxide detector for an additional $550. The new price of $4,050 will have a new monthly payment of $81. Imagine having cleaner air to breathe, increased peace of mind, and enhanced safety for your family for only $11 more a month. I know Mom will like it!

It's all in the presentation. When presented correctly, you can improve the number of jobs closed by up to 37% and offer homeowners what they want. And happy homeowners lead to higher referral rates for you.

Commercial Service and Replacement

Commercial service and replacement, unlike commercial construction, has the potential to be extremely profitable thanks to the popularity of commercial service agreements. Commercial service agreements are a win–win situation for all involved. Customers love them because having such an agreement helps them manage their own finances. They know how much is being paid for important HVAC service and they know who to call when something goes wrong. Commercial service agreements are ideal for HVAC companies, too, because the customer is paying for anywhere from one to five years of service in advance. This is money the company can count on because it has already been received.

Getting into the Market

Getting into the commercial service agreement market isn't at hard as you might think. Studies have shown that up to 60% of all commercial building owners are not happy with at least some aspects of their HVAC system. These companies are tired of hiring HVAC contractors who don't do a good job because they are too afraid to charge the necessary fees to perform top-quality work.

If you want to win the confidence of a commercial customer, show him or her how you your services are better than those of your competition. Highlight the fact that you use top-quality or OEM parts;

tell them about your workers' education, experience, and certifications; and even share what other customers you have worked for so the company can ask these customers if they are happy with your work.

You can also offer a comprehensive analysis of their HVAC system so they know what the trouble areas are and how you can help to improve them. You may be tempted to offer this analysis for free, but don't do it. Charge a fee that covers your transportation costs and your labor overhead and which also allows you to turn a profit on the job. This isn't a small, quick job where a single worker goes around and has a quick look at stuff. If the building is quite large, you may need to send a couple of technicians. Even a small building will take time to fully examine as you look over the HVAC system and all its components, take note of circumstances and conditions that are putting a strain on the system, check the ducts for leaks, and much more. Charge a fair price for your services while letting your clients know that they will be getting their money's worth. Gross margins for commercial service work should be in the 50% to 60% range, depending on the cost of the parts. Remember, the benchmark for a commercial service contract revenue is around $200,000 annually.

Chances are, a comprehensive HVAC checkup will find a few problems that need to be addressed. Common ones include HVAC units that are the wrong size for the building, dirty coils, leaks, missing system components, EPA violations, inappropriate maintenance frequencies, or even broken parts. Make a list of every single problem you find and bring it to your customer's attention.

If the client does not already have an HVAC service agreement with another company, you have the golden opportunity to draw up an agreement right there. There are several types of agreements you could offer, such as a preventative maintenance agreement, full labor agreement, and full coverage agreement.

Naturally, costs depend on how much is covered under the agreement; however, don't assume that a higher price tag on an agreement means that you will make a good net profit on it. Once again, you must know your overheads for each and every aspect of your business so you can draw up service agreements at the right price. A preventative maintenance agreement, for instance, may not cost much, but it can be profitable because the only costs for the job involve labor and fuel costs to get to and from the site along with some filters and belts.

Full labor agreements and full coverage agreements should be priced fairly. Remember that you are promising to deliver services and goods at a particular time and you will have to deliver these even if a customer's system breaks down on the weekend or during your peak season. You may need a part that you don't have in stock and you may discover that the company selling the part must order it next-day air, adding additional unexpected costs.

The idea of turning a profit on every single agreement you make is one that even some successful HVAC owners disagree with. Kenneth in Ohio runs a successful HVAC company and notes that he only breaks even on his service agreements. While this would seem to contradict everything I am telling you in this book, sometimes it really is best to secure more profitable future repair business by simply breaking even on the service agreement to lock in that future business. Kenneth notes that the customers he has service agreements with often order parts and services that are not covered under the agreement. He adds that he makes about three times the service agreement price from customers who need more than what that service agreement offers.

You can do the same; however, this isn't where you should start. You can't take on a brand-new customer and assume that you will make a certain amount of money off him or her. In fact, you should not count on making anything more than what is listed in the service agreement. Of course, your customer may pleasantly surprise you by asking for extra parts and services. On the other hand, that customer may have a tight budget and only stick to what is covered in the agreement. For this reason, I once again urge you to make sure your agreement covers your overhead and provides you with a net profit. You can make exceptions to this rule once you have worked for your customer for many years and know what to expect, but don't do it just to get a new customer. Offer quality at a quality price instead of simply trying to get what looks like a lucrative job contract.

You also need to calculate your profit margin on these agreements properly if you want to earn money rather than lose it. When you do this, the most important thing to get right is future costs, not just present ones. These agreements can last for years, so you can't base your prices on what your current overhead is. You need to look to the future to see how much you will need to charge later to remain in business.

Crunch the Numbers

To get commercial service agreements right, you need to calculate inflation into your costs. If you look at the current inflation rates, you will see that they are rising in the range of 1.50–2.50% each month. This means that prices will rise slightly over the course of many months to match the economy's inflation rate. You can either calculate these percentages into your service agreements or purchase the parts you think you will need in advance so that you can get them at today's lower prices. I don't recommend the latter course of action because you can never predict with 100% accuracy what you will need and when you will need it. What's more, shopping for parts in advance means that you must pay for storage for your inventory.

You will also need to calculate inflation rates when you consider what your employees' salaries will be years from now. You can't expect to pay them the exact same rate year after year—prices are steadily rising and good employees will want their salaries to steadily rise as well so they can make a reasonable living.

Bear in mind that a service agreement is not a one-time job. Such agreements can last for one, two, three, or even five years. If you don't get your overhead right, you will be paying for it for a long time. Your losses will add up quickly, especially if you have many service agreements with price quotes that make it impossible to turn a profit.

Chapter Summary

If you are trying to grow wealth in your business, you need a growth strategy that includes developing a strong and well-advertised service department. It should be clear by now that service breeds replacements. Fewer than 25% of consumers are considered pre-planners: people who know they need a new unit and start proactively shopping. The rest of equipment purchasers are reactionary, meaning their purchase originated from a demand service call. Obviously, then, having a strong and well-marketed service department is essential to long-term growth.

Adding service agreements to your offerings keeps existing customers in your database, not your competitors'. Over time, these customers will need repair service and then eventually new HVAC systems. Residential and commercial service agreements keep these revenues in your company.

Chapter 6
Managing Multiple Businesses

Chapter Overview

In this chapter, you will learn:

- You own more than one business
- How to departmentalize your company
- How departmentalizing your company affects pricing
- The direct expense method

Have you ever considered owning multiple businesses? Pretend for a minute that you owned an HVAC company, a bicycle shop, and a Subway sandwich store. Would you say these businesses had distinct differences? Of course you would! The HVAC company is labor intensive, while the bicycle shop would be asset intensive with large investment in inventories and high retail location expenses. The Subway shop has time-sensitive inventories that must be properly managed to reduce lost profits from spoiled food products.

As the owner of three very different companies, you would understand that each one would have to be managed differently. Using the same management and marketing strategies for each of the three companies would result in disaster, as would grouping these three companies as a single business and interchanging funds between them.

However, an important point that most HVAC company owners do not understand is that as an HVAC owner, you *do* own multiple companies, all operating under the umbrella of your company name.

In every HVAC business, there are at least two different companies: an installation company and a service company. They are different because of market cap competitiveness as well as differing overhead requirements.

Imagine selling a residential new construction job for a 60% gross margin. There is no way you would be competitive in the industry if you did this, as the market cap is currently no more than a 35% gross profit margin. At the same time, though, you would not want to set a limit of making no more than 35% gross profit margin on your parts; in fact, the market cap in the parts industry is currently a 60–70% gross profit margin.

Now take a moment to think about your service department. Your service department depends on vehicles to get technicians from one to job to another. If you have a lot of jobs, your technicians will be out in those vehicles all day. You will need to pay for fuel for the cars and you will need to pay your technicians for a full day's work even though they won't be performing service tasks eight hours a day. This means you will incur costs for unbillable time. What's more, this is not a small amount of time; in fact, your technician may spend up to a third of his or her day driving to or from service appointments. As you can see, the percentage of overhead for your service department will probably be higher than the overhead percentage for your parts replacement department.

These examples make it clear that each business sector needs to have different cost and overhead structures. Furthermore, each of these structures must be managed differently. The market cap for retail replacement installations provides higher gross margins than the market cap in the residential builder market. However, these higher margins come with a price tag in the form of higher overhead costs from sales commissions and promotional overhead costs.

Here are the four most important reasons to departmentalize your HVAC business:

- Departmentalizing enables you to determine and set competitive price levels based on market cap for each department. This move will help to maximize your profitability.

- Departmentalizing makes it possible to use the single-factor pricing formula to accurately determine net profits on any job.

- Departmentalizing allows you to determine the breakeven sales point by department.

- Furthermore, departmentalizing makes it possible to keep tabs on sales trends and ratios so you can successfully plan the future of your business.

If your company handles residential HVAC replacement, new residential construction, and residential service, then each of these lines of work should be separated into its own department on your financial statement. Moreover, a company that does a minimum of 25% commercial work should have a separate department for this as well.

Establishing Departments

Once you've made the decision to departmentalize, remember to limit yourself to a manageable number of departments in order to ensure you are getting meaningful information. Having too many departments can dilute the significance of the information you gather.

Some key criteria in establishing departments include the following:

- The sales revenue for each department you are creating must equal at least 15% of total sales. If your company is large, make sure that the new department generates at least 25% of total sales.

- Each business group should be recognized. Ideally, you should keep installation and service separate. However, you may not be able to do so if one portion of your company doesn't account for much revenue. For instance, a company that focuses primarily on service and doesn't handle many installation jobs may want to group residential service and installation in the same category.

- Target demographic can also play a role in your company department division. Dividing commercial work from residential work can be wise, especially if you do a lot of commercial installation and repair. Even so, the deciding factor in this decision should be how much money you make from each line of work. You don't need to make commercial work its own department unless 15–25% of your gross income comes in from this line of work. If you don't get enough commercial work to meet this standard, pair commercial service with residential service and join the commercial and residential installation departments together.

- Bear in mind that departments should not be fixed. Your business will change over the years and you may need to create more departments in the future. Keep an eye on where your gross finances are coming from to know when to create new departments or even merge existing ones.

After the departments have been established, you must determine a way to allocate the correct overhead to each so you can accurately see how much overhead it takes to operate any given department. This is extremely important, as overhead is a key component to market competitiveness and pricing accuracy, as we've seen.

Overhead Allocation Methods

There are several methods used to allocate overhead between departments. None of them is an exact science, but the most commonly used methods are the direct labor method and the direct expense allocation method.

Direct Labor

Many industry consultants recognize the direct labor method as the easiest and most balanced form of overhead allocation. Using this method, the percent of overhead allocation for each department is determined by that department's direct labor as a percentage of the company's overall direct labor.

The following exercise will provide clarity about how the methods work. In this example, we have a company that has three divisions: Existing Home Replacements, New Construction, and Residential Service. The direct labor for each department is listed below:

Replacement Labor: $86,800

New Construction Labor: $55,033

Service Labor: $68,632

The total sum for direct labor comes to $210,465.

To figure the correct amount of overhead to charge each department, you divide each department's direct labor by the total company direct labor to get the allocation percentage for each department.

To determine the allocation percentage factor, you divide the department labor by the total (consolidated) labor figure. If, for example, you divide the $86,800 listed above for replacement labor by the total $210,465 equals 41%. This means 41% of the company's overhead would be allocated to the replacement department.

Once the labor allocation percentage has been determined for each department, simply multiply each overhead expense item that is shared between the departments on your income statement. This will allow you to allocate the exact portion of the expense item to each department. Here's an example that makes this point clear:

ABC Heating and Air has a general office salary expense of $50,000. To determine each department's share of this expense, simply apply each department's labor factor to that $50,000.

> Replacement Department: $50,000 x 41% = $20,500
>
> New Construction Department: $50,000 x 26% = $13,000
>
> Service Department: $50,000 x 33% = $16,500

A lot of expenses will be shared between departments. Some examples include fuel and vehicle maintenance, the cost of office space for your business, storage costs, and the cost of office supplies.

Taxes are another expense that we haven't talked about much yet. This, too, should also be shared between all departments. Thankfully, like many of the overhead expenses we've outlined, it is an overhead that can be minimized. Hire a good accountant to look over your taxes and find any deductions you can take advantage of.

The Direct Expense Method

This method, when used in conjunction with the direct labor method, provides the highest degree of accuracy and fairness when allocating overhead between departments. Let's say you have a dedicated service manager who does not have any responsibility for any of the business activities in the installation department. It would not be accurate to charge some of the service manager's wages to the installation department. Overinflating a department's overhead with expenses that should be charged directly to a single department will result in overinflated prices in the overcharged department, which could price them out of the market when bidding jobs.

The main point here is that when expenses can be directly traced to a single department, those expenses should be charged to that department (and that department alone). All other expenses that are shared between departments should be allocated to each department proportionally using the direct labor method of expense redistribution.

Owner's Salary

Knowing where to draw the owner's salary from can be difficult, as the owner works with all business departments. There is no perfect way to calculate this; however, most experts advise taking the sales of each department and dividing them by the total sales of the company to determine the percentage of sales that each department generates. This would be used to determine the percentage of salary that each department would pay toward the owner's salary.

The reason that experts use this calculation method to determine the owner's salary allocation is that in most cases, the owner spends the most time with the departments that generate the most sales. This method should also be used when determining how to allocate salaries for managers who work with all company departments.

Let's use ABC Heating and Air an example yet again so you can see what I mean.

ABC has a total sales revenue of $1,000,000. The company is broken down into three divisions: Existing Home Replacements, New Construction, and Residential Services. About 50% of all sales come from Existing Home Replacement. Residential Services accounts for 35% of remaining sales, while 15% of sales come from New Home Construction. When allocating overhead for your own salary as business owner, you would take 50% of your salary from Existing Home Replacement sales, 35% from the Residential Services Department, and 15% from New Home Construction.

It may seem complicated to work out your salary in this way, but it's really the only accurate way to do it. Charging your salary to a single department will raise the overhead calculations for that department. This will, in turn, cause you to price jobs above the market cap because you think you have to raise a higher net profit from these jobs than is really necessary.

Advertising Expense Allocation

The majority of successful HVAC companies do a fair bit of advertising in order to generate more sales; however, most spend very little on advertising for the construction department. Given this fact, the most accurate way to allocate advertising expenses is to divide this overhead between the replacement and service departments.

Let's have a look at an example. Rightway HVAC spends about $10,000 on advertising every single year. The company should then divide this expense equally between the replacement and service departments.

This may seem inaccurate if you are spending most of your advertising dollars on promoting your replacement services instead of your service department. However, in reality, both of these departments are intertwined to a certain degree. In most cases, your service jobs will generate replacement work, as the customers who appreciate your service will turn to you when they need a new HVAC unit. The reverse is also true. If you do a good job installing an HVAC unit, the customer will call you back when they need service.

Sales Dollars per Advertising Method

We now jump to a complex sales subject: advertising. Remember that when we talked about overhead percentage in the first chapter of the book, I recommended a sales and advertising budget of 3–5%. I still recommend this advertising budget to any HVAC company owner who wants to generate sales and profits on a regular basis. New companies will likely be on the higher end of the scale, whereas companies with an established clientele can afford to pay a little less for advertising than they did in the past.

Now the key is to determine how much to expect from your advertising efforts. Experts note that the return on advertising investment ratio for a good company should be about 5:1. This means that you generate about $5 in revenue for each dollar you spend.

As an example, let's have a look at direct mail postcards, which are presently one of the most commonly used forms of advertising in the industry. They are estimated to have a direct return rate of about 1.5% because they are only effective in reaching people who need immediate HVAC repair or replacement. This means that you are bringing in about 10 customers for every 1,000 postcards you send. Now bear in mind that the average cost of sending each postcard comes to $0.47. This means that you are spending $470 just to send out postcards.

If this were a direct mail campaign targeting people who needed service the results would look like this. The average service ticket ranges between $250 to $350 nationally depending on your geographic area. This means the 10 leads you gained with this campaign would return gross service revenues between $2,500 and $3,500 or a minimum return of 5 dollars of revenue for every one dollar spent on advertising (5:1). It's fair to say that this is an effective form of advertising.

What's more, at least some of these customers will likely purchase goods and/or services from you more than once, thus exponentially increasing your return on advertising.

You should try to apply the same formula to every single form of advertising you spend money on, including radio ads, direct mail, and TV ads. An ROI that is slightly less than 5:1 is acceptable; however, anything significantly less is simply not worth it.

The same principle also applies to internet marketing. The internet has opened numerous advertising opportunities that you can take advantage of at little or no cost. Some of these have the potential to generate numerous sales. The problem is that you will either need to learn about internet marketing tactics such as content generation, SEO, and keyword usage or else outsource the job to a firm that is familiar with internet marketing strategies.

If you opt to outsource your internet marketing, apply the same principle to that firm that you would apply to a salesperson. Your chosen marketing firm should generate enough sales to justify the expense. If it can't do so, then cut the cost.

The most important thing to remember from this section is that you need to spend up to 5% on advertising and promotion. That is the only way you can afford to hire a salesperson and/or internet marketing firm. While you do need to make sure that the company or person you hire is making enough money for the company to justify the expense, don't hold back from getting outside help just because you are afraid of the cost.

Customer Demographics

Use your departmentalized income statements to define your mix of business and who those customers are. By doing so, you get a clearer picture that can show you if service is driving your replacement opportunities or if those sales originate through direct marketing and/or referrals for your installation business. You will then know what demographics to target when you create an advertising campaign. This data will also affect what type of marketing you do.

Will the marketing focus on service-type ads to expand your service business or replacement-type ads to grow that business?

Monetary Information

How much is the average homeowner customer in your area spending on HVAC repair? How much does the average business owner spend to have a new HVAC unit installed? Your departmentalized income statements will enable you to see what type of clientele you are reaching. For the service department, divide the total number of service invoice tickets by the service department's revenues. The average invoice ticket nationally is between $250 to $350 depending on geography. For your equipment replacement sales, take the number of installation jobs performed and divide by the installation revenues for the installation department. Average replacement system price for a 3-ton unit, 14-SEER air conditioner is $6,500–$7,000.

Chapter Summary

Once an income statement has been departmentalized, you and/or your management team can extract a wealth of information from your income expense sheet. Many owners mistakenly look at the bottom line and accept it for what it is, then file it away without performing any analysis. As it turns out, these owners miss a golden opportunity to uncover the true story behind why the net profit number looks the way it does.

By understanding how to analyze the income statement, the owner or management team can make decisions that may dramatically improve financial performance. The income statement is a treasure trove for finding out how well your operation generates net profit. Knowing the throughput ratios that measure this productivity can reveal how effective your company is at utilizing manpower. This knowledge helps identify weaknesses in service technician or installer revenue performance and allows you to set appropriate headcounts or staffing levels for each part of the company.

Additionally, we learned that market cap pricing is different for each type of work your company performs. Departmentalizing your company allows you to price each job accordingly to maximize the market cap for that segment of the business.

Scott Ritchey with Gary Kerns

Chapter 7
We Don't Know What We Don't Know: Financials

Chapter Overview

In this chapter, you will learn:
- Four things a contractor must know
- Cash vs. accrual accounting method
- What makes up an income statement
- What makes up a balance sheet

I am sure you have heard the phrase, "you don't know what you don't know." Please tell me what that means. Any time I have heard someone say that, they were using it for an excuse as to why they could not get something done. My first response is always, "Why didn't you ask?"

After spending decades consulting with HVAC contractors of all ages and walks of life, I have come to the following conclusion: No one likes to be told what to do!

When we were young, we were totally dependent on advice from our parents or people who were older than us. We listened and (at least sometimes) obeyed what they had to say because we knew they were wiser than us and had our best interests at heart.

As we grew older and more independent, we naturally wanted to make our own decisions. We started to resist being told what to do. This helped us grow up, as we learned to not only decide our future but also take responsibility for our actions and decisions.

We are all grown up now, but the desire to remain independent and not have people tell us what to do and how to do it remains strong. This holds especially true if the topic up for discussion is one that we know a lot about.

Unfortunately, pride remains one of the biggest obstacles when it comes to helping HVAC contractors turn their company around. Some of these contractors have been in the HVAC industry for decades. They have handled a wide range of projects. They have more field experience than I do. So it's not surprising that they are not all that willing to listen to what a business expert might say about their business. They are even less inclined to help the business expert give accurate advice by showing their financial statements to a total stranger—even one who is promising to help them make sense of these complicated reports if they will only allow them to be seen.

As it turns out, there are times when the stakes are too high *not* to tell someone what to do—situations where the consequences are severe, costly, and/or irreversible. I am not proud of an industry where 47% of contractors fail in the first four years of operation and 90% fail by year 10, all because someone did not like to be told what to do regarding their business practices or did not have the confidence to share their circumstances with a supplier or sales rep who might have been able to help them.

Most of the advice shared in this book is not given by suppliers who have failing contractor customers, because they themselves do not understand the contracting business. In rare instances where the supplier *does* understand the financial basics of running an HVAC business and sees that a customer is struggling and making serious mistakes, the supplier is often uncomfortable offering assistance that could help the struggling contractor. Conversely, it is sad to see a contractor who is naturally too proud to tell the supplier that there is serious trouble with the business and ask for help.

When I say, "I just want to make it easy for you," I really mean that. The information I am about to share is not opinion. It consists of foundational business truths that a contractor needs to understand and implement in order to survive and succeed in an industry that breeds such high rates of failure.

Four Things a Contractor Must Know and Do to Consistently Turn a Profit

Business finances can seem complicated if you aren't familiar with financial terminology and/or are running a business for the first time. However, the principles behind the terminology are not as complex as it might initially seem.

Here are the four most important things a contractor must know and do to make money. These points aren't listed in any particular order;

they are equally important, and missing just one of them could spell disaster for your business.

1. An HVAC business owner must have defined pricing and job cost evaluation procedures that are followed by all employees who are responsible for pricing work. A company with project managers must ensure that the managers are well versed in these principles and apply them to every potential job and job bid.

2. An HVAC business owner must understand that there are several business segments within an HVAC business. Each has its own overhead and pricing structure. The market cap for each segment dictates what gross margin percentages can be obtained; this in turn determines the pricing policy for the business segment.

3. An HVAC business owner must receive a set of monthly financials that includes an income statement (also known as a profit and loss statement) and a balance sheet. However, receiving them is just a start. The business owner must be able to read and understand what these invaluable pieces of paper are saying. So many poor business practices show up in the financials. It is not about the bottom line being black or red. It is about what *caused* the financials to be black or red. Knowing the reasons allows you to adjust your policies and procedures to improve financial performance.

4. An HVAC business owner must understand and perform three forms of interpretive analysis on the financial data found in the financial statements: balance sheet ratio analysis, performance benchmarking to industry data, and business throughput, which is the rate at which a system achieves its revenue goal through productivity measurements and benchmarks.

Since most of this book thus far has focused on pricing techniques and overhead discussions, let's start with the forms of analysis that can be used on the income statement and the information they provide.

The Income Statement

The income statement shows how much money a business has made or lost over a specific period of time. Income statements are usually generated monthly, quarterly, or annually. Because HVAC tends to be a risky business venture and it's easy to miss mistakes until it is too late, I strongly recommend that you print out and examine an income statement every single month, without fail. Generating a monthly analysis might seem like a lot of work, but once you learn the analytical techniques involved, the job can be completed in a couple of hours or less. That's not much time in the grand scheme of things, and it is time well spent if you want a competitive advantage over most other contractors.

The income statement is a summary of the company's revenues, costs, and expenses during a specified period of time—in this case, over one month. Financial performance is assessed by giving a summary of how the business incurs its revenues and expenses through both operating and non-operating activities.

There are two methods of reporting income statements. They are the Cash Basis method and the Accrual Basis method.

The Cash Method

The Cash Basis method of reporting an income statement recognizes transactions only when cash has been received or paid out. This method reflects checkbook accounting. The problem with using this method for a contracting business is that it does not clearly match costs to related sales transactions in the same income statement accounting cycle.

For instance, an HVAC company that uses service agreements is paid for these agreements in advance. If you manage to obtain two or three of them in one month, your income statement is going to look pretty good. You will have a large amount of income and no expenses to match them. However, down the line your income statement will look horrible when you have to actually spend money buying parts and sending technicians to repair the HVAC units that you have agreed to service.

This is just one example out of many. I believe the point is clear, though: the cash method of reporting an income statement is simply not good enough for an HVAC company. It could work in some instances if the company is generating an annual income statement instead of a monthly one, but it is far better to use the accrual basis method explained below.

The Accrual Method

The Accrual method recognizes revenues, costs, and expenses as realized, even if cash has not been exchanged. The reason this method is the preferred choice for HVAC contracting is because all transactions are posted in the period they occurred in.

The Income statement is broken down into five basic sections: Revenues, Cost of Goods Sold, Gross Profit, Overhead Expenses, and Net Profit. Each section is important, so let's look at each one in detail so you can better understand what your company's income statement is telling you.

The Revenue Section

This is the section where all types of sales should be grouped and recorded. At a minimum, sales categories for the HVAC business should include Installation Sales, Service Labor Sales, Service Parts Sales, and Service Agreement Sales. If 25% or more of the business is engaged in commercial sales activities, then revenue and cost of sales categories for that division should be set up as well.

Chapter 6 discussed all the reasons why you should departmentalize your business. It all starts by setting up the income statement correctly. Don't group all the sales together. Make sure each one has its own revenue sales line so you can see exactly where the sales are coming from for your business. This allows you to see where your marketing efforts must be concentrated if you want to grow one business segment over another.

Look at the right side of the Income Statement – Revenue Section graphic and you will see the KPI section. KPI stands for key performance indicator. These indicators serve as benchmarks against which to gauge your company's performance. Being within the range of the benchmark indicators means your company is performing to the recommended industry ranges.

After years of consulting, it is our opinion that the most balanced companies have a 65% to 35% split between installation department sales and service department sales, particularly if the majority of the installation sales are in the equipment replacement segment. This is because equipment replacement sales and service sales are the segments that produce the highest gross margins. The higher the gross margins, the better chance you have for double-digit net profits as long as you are controlling your overhead costs.

Income Statement - Revenue Section			
Revenues:		% of Sales	KPI
Installation Sales	$610,000	73%	65% - 75%
Service Labor Sales	$128,534	15%	15% - 20%
Service Parts Sales	$64,275	5%	5% - 10%
Service Agreement Sales	$35,500	4%	3% - 5%
Total Revenues	$838,309	100%	

The Cost of Sales Section

As the name implies, the cost of sales section records all the "direct costs" incurred when your company does a job or performs a service task. Some accounts that you will find in this section include Equipment Costs, Supplies Costs, Labor Costs, Parts Costs, Permit Costs, Warranty Costs, and Subcontract Costs. If your company works with residential or commercial construction companies, you may need to add a column for Equipment Rental Costs. Costs incurred in this area may include renting a crane or hydraulic lift to handle a heavy-duty HVAC installation job.

Again, we want to departmentalize the cost of sales section so that we can use the KPIs to tell how well we are performing. Top-performing companies have installation cost of sales in the 50% to 55% range, yielding installation gross margins of 45–50%. With the service department, cost of sales should range between 35% to 40%, yielding service gross margins of 60–65%.

Income Statement - Cost of Sales Section			
Cost of Sales:		% of Sales	KPI
Installation Department			
Installation Equipment	$244,000	40%	38% - 42%
Installation Supplies	$30,400	5%	5% - 7%
Installation Labor	$50,000	8%	8% - 10%
Permits and Fees	$5,960	1%	1% <
Warranty Reserve	$7,880	1%	3% <
Total Installation Costs	$330,240	54%	50% - 55%
Service Department			
Service Labor Cost	$44,844	35%	30% - 35%
Service Parts Cost	$25,710	40%	38% - 45%
Call Back Labor	$2,562	2%	2% <
Warranty Labor	$3,844	3%	3% <
Total Service Costs	$79,960	35%	35% - 40%

The Gross Profit Section

Remember, gross profit is not net profit. It is the money you make from a job after you have covered all the job expenses. It is the money you need to cover overhead expenses such as insurance, office rental, advertising, your own salary, and vehicle maintenance and repair. Top-performing companies that have a replacement/service business model see gross profit margins between 48% and 52%. You would not see these types of gross margins in a company with a heavy new construction or commercial business model. Depending on the new construction mix between tract homes and custom homes, you would most likely see gross margins between 32–38%. Design build commercial contractors usually have gross margins between 38–42%.

Income Statement - Gross Profit Section

Gross Profit:		% of Sales	KPI
Installation Gross Profit	$279,760	46%	45% - 50%
Service Gross Profit	$148,349	65%	60% - 65%
Total Gross Profit	$428,109	51%	48% - 52%

The Overhead Expense Section

The Overhead section lists all the expenses that it takes to operate your business which are not directly related to performing a job. In Chapter 3, we covered the six categories of overhead. This Income Statement – Overhead Section graphic breaks down the overhead into these six categories and lists the KPIs for each type of expense.

Breaking down the Overhead section into these six categories provides a detailed look at your expenses. You can clearly see whether your overhead costs are in line with the industry benchmarks or not. This allows you to focus on costs that are out of line and need attention.

When I look at a client's income statement, I look at all areas but pay particular attention to the two areas that threaten net profit the most: payroll overhead and insurance overhead.

Since payroll overhead usually makes up half of the total overhead, it is critical to keep it in line with the benchmark. As for insurance, skyrocketing increases in health insurance premiums makes this a challenging benefit to offer your employees. I am not advocating not offering health insurance, but you must find ways to control costs in this area. To keep costs in the benchmark range, you will have to consider options like higher deductibles or doctor hotlines that handle simple health issues over the phone for lower costs than a doctor's office visit.

Controlling overhead costs is just as critical as knowing margin market caps and pricing to ensure maximum profitability for your company.

MAKE MORE MONEY: 12 PROFIT PILLARS FOR HVAC CONTRACTOR SUCCESS

Income Statement - Overhead Section			
Operational Overhead		% of Sales	KPI
Office Supplies	$1,839	0.2%	< .5%
Professional Fees & Licenses	$2,519	0.3%	< .5%
Small Tools	$1,939	0.2%	< .5%
Communications	$3,756	0.4%	< .5%
Office Equipment	$2,960	0.4%	< .5%
Unapplied Labor	$5,790	0.7%	< 2%
Bad Debts	$2,088	0.2%	< .5%
Interest	$1,771	0.2%	< .5%
Depreciation	$4,658	0.6%	< 2%
Dues and Subscriptions	$855	0.1%	< .5%
Uniforms	$1,245	0.1%	< .5%
Total Operational Overhead	$29,420	3.5%	< 3%
Promotional Overhead			
Advertising	$6,719	0.8%	3% -5%
Vehicle Overhead			
Auto - Fuel, Oil, Tires	$10,498	1.3%	< 1.5%
Auto - Repairs	$6,677	0.8%	< 1.5%
Total Vehicle Overhead	$17,175	2.0%	< 3%
Payroll Overhead			
Owner Wages	$65,000	7.8%	6% - 8%
Office Wages	$99,486	11.9%	10% - 12%
Payroll Taxes	$19,244	2.3%	2% - 3%
Total Payroll Overhead	$183,730	21.9%	< 22%
Insurance Overhead			
Health Insurance	$38,334	4.6%	< 5%
Workers Compensation	$13,455	1.6%	< 2%
General Liability & Auto	$8,275	1.0%	< 1.5%
Total Insurance Overhead	$60,064	7.2%	5% - 7%
Facility Overhead			
Rent	$12,000	1.4%	< 1.5%
Utilites	$3,277	0.4%	< .5%
Total Facility Overhead	$15,277	1.8%	< 2%
Total Overhead	$312,385	37.3%	37% - 40%

The Net Profit Section

The net profit section is the final "judge and jury" part of the income statement—it shows you how well the company is performing. To arrive at the net income number before taxes and interest, you subtract total overhead dollars from your total gross profit dollars. We all know the goal is to have a positive number, but if you find yourself in the red, the good news is that there are plenty of ideas in this book to get you back to black.

The acronym that I put in parenthesis beside the words "net profit" in the graphic is a financial term used by accountants. EBITA stands for "earnings before taxes, interest, and amortization." Your true net profit would be the remaining profits after the interest, taxes, and amortization adjustments have been made.

Net Profit (EBITA)	$115,724	13.8%	>7%

Putting It All Together: The Contractor Income Statement

		% of Sales	KPI
Revenues:			
Installation Sales	$610,000	73%	65% - 75%
Service Labor Sales	$128,534	15%	15% - 20%
Service Parts Sales	$64,275	5%	5% - 10%
Service Agreement Sales	$35,500	4%	3% - 5%
Total Revenues	**$838,309**	**100%**	
Cost of Sales:		% of Sales	KPI
Installation Department			
Installation Equipment	$244,000	40%	38% - 42%
Installation Supplies	$30,400	5%	5% - 7%
Installation Labor	$50,000	8%	8% - 10%
Permits and Fees	$5,960	1%	< 1%
Warranty Reserve	$7,880	1%	< 3%
Total Installation Costs	**$330,240**	**54%**	**50% - 55%**
Service Department			
Service Labor Cost	$44,844	35%	30% - 35%
Service Parts Cost	$25,710	40%	38% - 45%
Call Back Labor	$2,562	2%	< 2%
Warranty Labor	$3,844	3%	< 3%
Total Service Costs	**$79,960**	**35%**	**35% - 40%**
Gross Profit:		% of Sales	KPI
Installation Gross Profit	$279,760	46%	45% - 50%
Service Gross Profit	$148,349	65%	60% - 65%
Total Gross Profit	**$428,109**	**51%**	**48% - 52%**
Operational Overhead		% of Sales	KPI
Office Supplies	$1,839	0.2%	< .5%
Professional Fees & Licenses	$2,519	0.3%	< .5%
Small Tools	$1,939	0.2%	< .5%
Communications	$3,756	0.4%	< .5%
Office Equipment	$2,960	0.4%	< .5%
Unapplied Labor	$5,790	0.7%	< 2%
Bad Debts	$2,088	0.2%	< .5%
Interest	$1,771	0.2%	< .5%
Depreciation	$4,658	0.6%	< 2%
Dues and Subscriptions	$855	0.1%	< .5%
Uniforms	$1,245	0.1%	< .5%
Total Operational Overhead	**$29,420**	**3.5%**	**< 3%**
Promotional Overhead			
Advertising	$6,719	0.8%	3% - 5%
Vehicle Overhead			
Auto - Fuel, Oil, Tires	$10,498	1.3%	< 1.5%
Auto - Repairs	$6,677	0.8%	< 1.5%
Total Vehicle Overhead	**$17,175**	**2.0%**	**< 3%**
Payroll Overhead			
Owner Wages	$65,000	7.8%	6% - 8%
Office Wages	$99,486	11.9%	10% - 12%
Payroll Taxes	$19,244	2.3%	2% - 3%
Total Payroll Overhead	**$183,730**	**21.9%**	**< 22%**
Insurance Overhead			
Health Insurance	$38,334	4.6%	< 5%
Workers Compensation	$13,455	1.6%	< 2%
General Liability & Auto	$8,275	1.0%	< 1.5%
Total Insurance Overhead	**$60,064**	**7.2%**	**5% - 7%**
Facility Overhead			
Rent	$12,000	1.4%	< 1.5%
Utilities	$3,277	0.4%	< .5%
Total Facility Overhead	**$15,277**	**1.8%**	**< 2%**
Total Overhead	**$312,385**	**37.3%**	**37% - 40%**
Net Profit (EBITA)	**$115,724**	**13.8%**	**> 7%**

Scott Ritchey with Gary Kerns

The Balance Sheet

The balance sheet is the financial statement that reports the financial condition of the company at a specific point in time. The values are always changing with cumulative impact, and they can either change positively or negatively. The accounts tied to the balance sheet never reset to zero.

The balance sheet is the financial statement that shows us what we own and what we owe. It is the document that bankers and credit departments scrutinize to determine the overall financial health of a company.

There are three key sections of the balance sheet: the Asset accounts, the Liabilities accounts, and the Equity accounts. Assets are what we own. The liabilities are what we owe, and the equity section is what we are worth in terms of book value. The balance sheet tells a story of how well we managed our asset resources in creating our wealth and the company's value.

The importance of having a strong balance sheet is clearly demonstrated if you ever need to get a line of credit to ease cash flow requirements at a particular time when you have less income and lots of expenses. Such times happen periodically in the HVAC industry, so don't be surprised when you hit this roadblock. A strong balance sheet is also extremely advantageous when you need to find additional suppliers because of credit limit constraints with existing suppliers.

The balance sheet acts as a compass in controlling the growth rate of the company through equity management, protecting the company from becoming over-leveraged by debt. It shows the liquidity of whether the company can service debt and accounts payable as the company navigates the seasonality of the business.

It is not a coincidence that the balance sheet was named that. As mentioned before, the balance sheet tells us what we own, what we owe, and what we are worth. Assets are those items we own, which are often financed by a bank or through our suppliers with lines of credit, creating liabilities owed, with the difference between the two being our equity or net worth.

The balance sheet equation helps us to know that the financial data on the balance sheet indeed balances. The equation is:

Total Assets = Total Liabilities + Equity

You need to review your balance sheet before any ratio analysis can be done. At this time, you must make sure that the total asset figure is equal to the total liabilities figure plus the equity figure. If they do not match, there has been an accounting error somewhere. This error must be corrected before going any further.

The Balance Sheet

Now let's take a deeper look at each of the parts of the balance sheet.

The Asset Section

Remember: asset accounts are items the company controls and utilizes. Basic accounts assigned to the Asset section include Cash, Accounts Receivable, Inventory, and Prepaid Expenses. These are classified as Current Assets because they can be converted to cash within a one-year period or the normal annual operating cycle of the business.

Every HVAC business owner knows what cash in hand is. However, the other items may not be quite as clear. Let's have a closer look at them:

Balance Sheet - Current Assets Section

Current Assets	
Cash	$83,006
Cash -Payroll	$10,000
Accounts Receivable	$96,916
Inventory	$19,920
Pre-Paid Expenses	$9,515
Total Current Assets	$219,357

Accounts Receivable

Accounts receivable is a term denoting money that you will be paid for work that you have already done or are doing right now. It is money that people owe you, but it is not really debt because it is money that you weren't expecting to receive right now.

Inventory

Inventory denotes items that you have in stock. However, be aware that your inventory isn't necessarily the same thing as your assets. Assets are items you own outright, and a good part of your inventory likely falls under this category. If, on the other hand, you are still paying on time for items in your inventory, then these items aren't really assets.

Prepaid Expenses

You will often have to pay certain expenses up front. Insurance is one glaring example of such an expense, as is office or storage room rental. You will probably take out an insurance policy for at least a year and pay the full cost of it at one time. However, this prepaid expense shows up positively on your balance sheet because it is not the same as paying for an HVAC part for a particular job. The prepaid expense saves you from having to pay a certain amount of money for the rest of the year; furthermore, the money you have paid upfront guarantees you a certain service or product for the full extent of the payment period.

Non-Current Assets

Better known as fixed assets, non-current assets are items that cannot be readily turned into cash and are not consumed within a normal yearly operating cycle. Examples of fixed assets include the vehicles you own, office furniture and equipment, machinery and equipment you own and use for your HVAC work, land and buildings, and accumulated depreciation. The main characteristic of fixed assets is that they are held primarily for use by the organization and are not usually offered for sale unless the business is in dire financial straits. Moreover, fixed assets tend to have relatively long usage life cycles.

Another account in this section of the balance sheet to be aware of is the accumulated depreciation account, which is listed with the fixed assets. This account is known as a contra account to the other fixed asset accounts. When a fixed asset is first purchased, it is recorded in the balance sheet at the value it cost to acquire it. Over time, the asset loses its original value and the IRS allows you to write off this expense in the form of depreciation.

The accumulated depreciation account represents the amount of depreciation expense recorded over the depreciated life of the item. This figure is often accounted for at the end of the year by your accountant. The reason the accumulated depreciation account has a negative balance is because it is the offset account for the depreciation expense account on the income statement and reflects the lost value of the asset.

Balance Sheet - Fixed Assets Section	
Fixed Assets	
Autos & Trucks	$134,653
Office Furniture & Equipment	$36,665
Machinery & Equipment	$86,317
Land & Buildings	$284,000
Accumulated Depreciation	($396,339)
Total Fixed Assets	$145,396
Total Assets	$364,653

The Liabilities Section

Liabilities are realized debts incurred through normal operational activities. A liability is defined as money owed for an obligation of indebtedness.

Like assets, liabilities are also categorized into current and long-term time frames for accounting purposes.

Current Liabilities

Current liabilities are debts incurred from normal business transactions done on credit or obligations accrued requiring future payment that will be repaid during the next 12 months. Examples include Accounts Payable, Accrued Payroll Taxes, Bank Lines of Credit, and the Current Portion of Long-Term Debts.

Balance Sheet - Current Liabilities Section	
Current Liabilities	
Accounts Payable	$52,658
Accrued Payroll & Taxes	$10,000
Short-Term Notes Payable	$6,916
Bank Line of Credit	$20,000
Other Current Liabilities	$18,811
Total Current Liabilities	$108,385

Accounts Payable

Accounts payable are the exact opposite of accounts receivable. These are accounts that you owe money to. Accounts payable are often owed to HVAC parts and equipment suppliers, companies that service your vehicles, and companies that you rent equipment from. These are accounts that you need to pay within the business year if you don't want to land in financial trouble.

Accrued Payroll Taxes

Taxes are, unfortunately, part and parcel of running any business. Accrued payroll taxes are taxes that you need to pay on your payroll but haven't paid yet. You aren't meant to pay them until the tax season; however, these taxes are still a liability because it is money you will owe to your state or federal government. When you file your quarterly 941 tax forms and pay the taxes, these accrued expenses move from the balance sheet to the income statement and show up in your overhead expenses as payroll taxes.

Bank Lines of Credit

Having a bank line of credit is a good thing. As we touched on above, HVAC work is seasonal. You may have enough jobs every single year to remain successful; however, you probably won't have the same number of jobs every single month. There will be busy months and months that aren't very busy. Having a bank line of credit enables you to stay in business by providing you with the money you need when work isn't plentiful. If you price your jobs right, you should have no problem paying the bank back when you do get work—even so, the bank lines of credit aren't given for free. They cost money, and the money you must pay to keep the line of credit going is a liability.

Long-Term Liabilities

Long-term liability debt is typically debt financed for a two-year period or longer. Examples include Notes Payable (Auto), Mortgage Notes, Notes Payable (Owner) and Less Current Portion of Long-Term Debt. This section of the balance sheet shows your long-term obligations to institutions you have borrowed money from to fund the company for the future.

Balance Sheet - Long-Term Liabilities Section

Long-Term Liabilities	
Auto Notes Payable	$28,500
Equipment Notes Payable	$10,000
Building Loans Payable	$74,515
Total Long-Term Liabilities	$113,015
Total Liabilities	$221,400

The Equity Section

The Equity section of the balance sheet represents what you've invested in the business. Once again, it breaks down into several parts—here, we'll take a look at each.

Owner's Equity

Owner's equity represents the book value worth of the company. It is the dollar balance remaining from Total Assets minus Total Liabilities. There are four basic accounts in this section. These include capital stock, owner's withdrawal, current earnings, and retained earnings.

Balance Sheet - Equity Section	
Equity	
Capital Stock	$1,500
Retained Earnings	$141,753
Total Equity	$143,253
Total Liabilities & Equity	$364,653

Capital Stock

Capital stock represents monies or assets that were contributed by stockholders to be used as the financial backing to start a corporation. This number is usually determined by a CPA in the first year of the business and remains unchanged.

Owner's Withdrawal/Dividend

These are disbursements made to shareholders after tax obligations have been met. The stockholders' withdrawals or dividends have a negative impact on the net worth of the company, as they are disbursed as income while showing up as an expense on the income statement. Retaining net profit in an organization improves long-term working capital requirements to fund future growth.

Current Earnings

Current earnings represent the sum of the net profit that the company has generated in the current fiscal year. Some CPAs like to record this separately from the retained earnings number to show the owner the current year's earnings, to which taxes will be applied. Because retained earnings reflect an accumulation of reinvested earnings over time, taxes have already been paid on those earnings. Showing a current earnings amount on the balance sheet is not required for accounting purposes.

Retained Earnings

Retained earnings are the cumulative sum of net profit generated by the company since its startup date. Retained earnings do not include the current earnings from the current fiscal year, and any withdrawals or dividends paid to stockholders are subtracted from this figure. Retained earnings are reinvested capital to help fund future company growth opportunities.

The information provided by the balance sheet is essential to understanding how solvent the business is in good times and bad. It also shows how well your assets are managed to create profitability and provide a return on investment into the business. If the return on business investment is not greater than the average rate of return on investment that the stock exchange provides, then you would be better off liquidating the company and investing the proceeds in the stock market. After all, why work so hard if the only result from your efforts is that you are breaking even?

Chapter Summary

Okay, I hear you. Reading what makes up a financial statement can be very dry, if not utterly boring. But you *must* be able to recognize each element of the income statement and balance sheet to be able to perform financial analysis and benchmarking. These tools can show you the financial health of your company as well as point out pricing, margin, and cost problems. Once you learn how to perform ratio analysis, you will always want to know what your financials are telling you.

Scott Ritchey with Gary Kerns

Chapter 8
Ratios: What Your Financials Tell You

Chapter Overview

In this chapter, you will learn:
- Solvency ratios: How stable are you?
- Working capital: What it is and how much you need
- Return ratios: Is your company a good investment?
- Productivity ratios: Do you run an efficient business?
- How to use benchmarking as a measurement tool

"Ratios" is a financial term that most people don't find very appealing. However, using ratios is a very good way to determine the financial health of your company and the concept isn't all that hard to grasp.

Ratio analysis is the most effective way to measure the utilization of the company's assets. It is also used to determine the organization's financial health and performance. Ratios used to analyze the balance sheet are broken down into two groups: Solvency Ratios and Profitability Ratios (also known as Asset Management Ratios).

The Importance of Solvency

Solvency ratios show us how liquid an organization is. They demonstrate the company's ability to weather any unforeseen changes in the usual ebb and flow of the business. Using solvency ratio analysis allows management to clearly see the company's ability to pay its current obligations. These ratios also highlight the amount of equity the ownership has in the business compared to borrowed debt from investors, bankers, or suppliers. In addition to liquidity, solvency ratios reveal how much working capital the company must have to fund future growth goals.

Here, we will focus our analysis on four key solvency ratios: the Current Ratio, the Quick Ratio (also known as the Acid Test), the Debt-to-Equity Ratio, and the Working Capital Ratio.

Solvency Ratio Calculation Formulas	
Current Ratio	$\dfrac{\text{Current Assets}}{\text{Current Liabilities}}$
Quick Ratio	$\dfrac{\text{Cash + Receivables}}{\text{Current Liabilities}}$
Debt To Equity	$\dfrac{\text{Total Liabilities}}{\text{Total Equity}}$
Working Capital	$\dfrac{\text{Current Assets - Current Liabilites}}{\text{Total Sales}}$

Current Ratio

This ratio is used to measure the company's ability to meet current debts as they come due. It measures how many dollars of current assets are available to pay current liabilities owed. The preferred ratio is 2:1. This means there are $2 in current assets for every $1 of liabilities owed.

The current average for the HVAC industry is 1.88. Simply put, the average HVAC contractor has $1.88 in current assets for every $1 in current liabilities. To ensure financial stability, contractors should have a goal of at least 2:1 for a current ratio.

$$\text{Current Ratio} = \dfrac{\text{Current Assets}}{\text{Current Liabilities}}$$

Quick Ratio

The Quick Ratio is a more stringent test than the current ratio for stating the company's liquidity position and its ability to meet current debts as they come due. This formula excludes inventory that could be sold to help generate cash to pay off debts. It considers only the company's cash and accounts receivable to measure its ability to pay current debts owed.

The current industry average is 1.59:1. This means that the average HVAC contractor has $1.59 in cash and receivables for every dollar in current liabilities. The minimum ratio is 1:1, which means for every dollar in cash and receivables, there is one dollar in current debt owed. Anything less than a 1:1 ratio indicates that the company is insolvent and in serious financial trouble.

$$\text{Quick Ratio} = \frac{\text{Cash + Receivables}}{\text{Current Liabilities}}$$

Debt to Equity Ratio

This ratio measures how much equity is in the company compared to how much debt has been borrowed or accrued. As a rule, this ratio should never be more than 1:1. Anything higher than that means the business is over-leveraged and the creditors have more ownership in the business than the owner does. However, if you discover your debt to equity ratio is higher than 1:1, don't panic. Sometimes it is necessary to over-leverage your business to expand it. Still, be aware that borrowing for expansion is only a viable option if you can generate adequate from the new business to meet the debt obligation—and additional cash flow to justify the expansion.

Don't tolerate a ratio of 1:1 for very long. Know what you are doing and have a plan in place for paying your creditors. Failure to do so could result in having to declare bankruptcy, a decision that could make it hard for you to start another business in the future.

$$\text{Debt To Equity} = \frac{\text{Total Liabilities}}{\text{Total Equity}}$$

Working Capital Ratio

This solvency ratio is recommended for many reasons. First, it should be used just to gauge if the company has adequate working capital to run the current business sales volume. Another use is in determining how much working capital would be needed to fund the business if it were to adopt an aggressive growth strategy like taking on a large multi-family project, a new subdivision with a new builder, or an aggressive internet marketing program to grow the replacement business by 25%.

A recommended target is 10% of total sales. When you are trying to forecast how much working capital you need for the upcoming year, you'll use next year's forecasted sales or the new projected sales revenue in the formula above. Businesses should strive to fund expansion efforts on their own instead of always relying on borrowed capital. The benefit of using your own money to expand your business is that you can save thousands of dollars that you would have had to pay in interest rates.

$$\text{Working Capital} = \frac{\text{Current Assets} - \text{Current Liabilites}}{\text{Total Sales}}$$

Manage for High Returns

The next set of ratios is referred to as asset management ratios or profitability ratios. Because they measure profitability through asset management, the information needed for each ratio comes from the income statement and balance sheet.

Profitability ratios measure how effective we are at using our assets to create profits and returns on invested capital. Ratios in this group are Return on Assets, Return on Sales, and Return on Investment.

Profitability Ratio Calculation Formulas	
Return on Assets	$\dfrac{\text{EBITA}}{\text{Total Assets}}$
Return on Investment	$\dfrac{\text{EBITA}}{\text{Total Equity}}$
Return on Sales	$\dfrac{\text{Total Sales}}{\text{Total Assets}}$

Return on Assets

This ratio determines how well assets are used to generate a rate of profit return. It measures profitability and management's effectiveness in asset allocation. This ratio uses EBITA, which is your net profit before interest expenses, taxes, and amortization are deducted from it. The net profit number will be found at the bottom of the income statement. The other half of the equation uses the total assets number found in the first section of the balance sheet. The current industry average return on assets is 49% for businesses with sales of $500,000 to $1,500,000.

Return on Assets	$\dfrac{\text{EBITA}}{\text{Total Assets}}$

Scott Ritchey with Gary Kerns

Return on Investment

Return on investment (ROI) measures the rate of return generated from the profits of the business to the amount of equity invested or retained in the business. To understand it better, consider the following question: If someone was investing in your business, what would be their rate of return? This is the true measurement of performance in the organization, and it's a key indictator used in business valuations when an owner wants to sell. Any number north of 33% return is a positive. The current return on investment for businesses with sales of $500,000 to $1,500,000 over the last five years averages 88%.

Return on Investment	$\dfrac{\text{EBITA}}{\text{Total Equity}}$

Return on Sales

This ratio measures how effective assets are at generating sales dollars for the company. The proper ratio to aim for here is 2:1 or 3:1. Put in simple terms, the goal is to generate between $2 and $3 for every single dollar invested in getting sales. Total sales will be found in the top section of the income statement while total assets are in the top section of the balance sheet.

Return on Sales	$\dfrac{\text{Total Sales}}{\text{Total Assets}}$

Industry Benchmarks: Solvency and Profitability

Benchmarks provide industry standards you can use when comparing your individual company performance to industry guidelines. The table below can be used to compare your solvency and profitability ratios to the industry benchmarks.

MAKE MORE MONEY: 12 PROFIT PILLARS FOR HVAC CONTRACTOR SUCCESS

Solvency Ratio Calculation Formulas		Industry Benchmark	Your Numbers
Current Ratio	Current Assets / Current Liabilities	2 : 1	___
Quick Ratio	Cash + Receivables / Current Liabilities	1 : 1	___
Debt To Equity	Total Liabilities / Total Equity	1 : 1 or 1<	___
Working Capital	Current Assets - Current Liabilites / Total Sales	10%+	___

Profitability Ratio Calculation Formulas		Industry Benchmark	Your Numbers
Return on Assets	EBITA / Total Assets	45%+	___
Return on Investment	EBITA / Total Equity	33%+	___
Return on Sales	Total Sales / Total Assets	$3+ Sales to $1 of Assets	___
Gross Profit Margin %	Gross Profit $ / Total Sales	47%+	___
Net Profit Margin %	EBITA / Total Sales	7%+	___

Productivity Analysis: Manpower Performance

Productivity analysis, sometimes referred to as throughput, measures how effectively we manage our human resources. By establishing productivity benchmarks, we can measure workforce capacity with efficiency outputs. Simply put, these benchmarks help determine the rate at which your company generates money through sales. Productivity analysis gets its source data from the income statement.

Because what we really sell is man hours, it is management's job to maximize the use of manpower. To do this, manpower ratios have been created to measure the effectiveness of labor generating sales dollars. However, bear in mind that these manpower ratios are merely benchmarks that have been created using industry averages. They are not absolutes.

Manpower ratios are broken into two groups: sales-related ratios and overhead-related ratios.

Scott Ritchey with Gary Kerns

Man-Power Performance Benchmarks

Sales $ Per Service Truck

Replacement Sales $ Per Installer

New Construction Sales $ Per Installer

Sales $ Per Employee

Sales $ Per Replacement Sales Person

Sales-Related Manpower Ratios

Sales Dollars per Service Truck

This ratio measures the amount of sales produced by one service technician and his or her truck. This ratio is often referred to as Sales Dollars per Service Truck. The benchmark rate is $150,000 to $200,000 per truck for a residential service truck and $200,000 to $250,000 for a commercial service truck. The benchmark uses an average service ticket rate of $219. Although a demand service ticket may average $335, technicians are not always running demand service calls. You must figure in the warranty calls, call-backs, and service maintenance tickets from service agreements into the total ticket count. Doing this gives you an average ticket of around $219. A service technician working for a $1,000,000 business runs about 800 to 850 calls per year. Doing the math at 825 calls would give us 825 x $219 = $180,675. Company location does play a factor in the call count and may also involve paying higher wages, so locations like Los Angeles, Phoenix, Chicago, and New York would have service truck revenues between $250,000 to $300,000.

$$\text{Sales \$ Per Service Truck} = \frac{\text{Total Service Sales}}{\text{Total Number of Trucks}}$$

Sales Dollars per Installation Technician

This ratio measures the amount of sales produced by one installer. Because some contractors may use two- or three-man crews per truck, you cannot use this as a per-truck measurement. As there are two types of residential installations, a separate ratio should be used for residential existing home replacements and residential new home construction installations. That's because the rate of production will be completely different for each. Replacement installations are usually completed in a day, whereas new construction installations for a typical 1,500-square-foot home with a 2.5-ton system will take up to four days to finish.

$$\text{Replacement Installation Sales} = \frac{\text{Total Replacement Sales}}{\text{Total Number of Installers}}$$

Roughly 80–85% of contractor businesses have revenues ranging between $500,000 and $1,000,000. Considering a residential replacement contractor, when you blend together complete system sales with equipment component sales like an air conditioner plus a coil or a furnace and a coil, the industry average per replacement sale is about $6,500. Adjusted for seasonality, a contractor with this profile averages about two installations per week (50 weeks), or 98 installs per year. This would generate $637,000 in replacement equipment sales ($6,500 x 98).

In this example, using a two-man crew (which is very common for the industry), each installer would be responsible for generating $318,000 annually ($637,000 / 2). Therefore, a range of $275,000 to $325,000 per installer is the norm.

This does not mean you can't have a higher revenue number per installer. For example, a million-dollar contractor whose business consists of 80% installation and 20% service can install up to 125 systems per year for a sales volume of $800,000. This would result in an installer revenue rate of $400,000 per installer.

Although possible, several things must come together to achieve this rate. One is a strong lead-generating program with a high-closing retail salesperson. Another is a good conversion rate, turning 15–20% of service calls into replacement sales—that quiet market share conversion we discussed briefly earlier. This concept is fully addressed in the service tech goldmine section in Chapter 9.

For a residential new construction contractor in the same profile range of $500,000 to $1,000,000 in sales revenue, the sales dollars per installer are much lower. The typical installation crew in the new construction market can do 1.5 homes per week based on the typical 1,500-square-foot home. The industry average price for a contractor for this type of home is $6,800. Annual revenues under this scenario would be 1.5 homes x $6,800 x 50 weeks = $510,000. If a two-man crew were used, then you'd be generating revenue of $255,000 per installer. In many cases, however, three-man crews are used in which a go-fer helper gets material, cuts holes, or retrieves tools. In this scenario, the revenue dollars per installer are now $510,000 / 3, or $170,000. Because these two methods of installation are often intertwined in the industry, a revenue per installer of $225,000 is the new-construction benchmark.

New Construction Sales	$\dfrac{\text{Total New Construction Sales}}{\text{Total Number of Installers}}$

Sales Dollars Generated per Employee

This ratio measures the number of sales dollars generated by the entire organization compared to the number of people who work there. The benchmark for this ratio is $140,000 to $175,000 per team member. The higher the number in this range, the better—but I do caution having a number higher than $175,000. Using data from BizMiner, the average gross margin percentage for a $500K to $1M HVAC contractor is 48%. Using the $150,000 per employee number, we see that an employee contributes $72,000 of gross margin. Is this enough to cover the company's overhead and still make a profit?

$$\text{Sales Per Employee} = \frac{\text{Total Company Sales}}{\text{Total Number of Employees}}$$

The answer is yes! According to BizMiner data, the average overhead for this contractor would be 40%. Taking 40% of $150,000 gives us an overhead number of $60,000. As you can see, the $72,000 gross margin covers the $60,000 of overhead and leaves a $12,000 net profit—that's an 8% net profit.

The reason I caution you on surpassing the $150,000 number is that over-burdening your employees with too much work will sap productivity. You'll quickly move into the realm of diminishing returns. When we reach this point, we are tasking the employees with too much work, making room for higher error rates, poor installation quality, poor service, increased overtime wages both for jobsite workers and office staff, and higher rates of employee turnover due to these conditions. So you can see that when used properly, this ratio can help you make better, more rational decisions about adding new employees.

Sales Dollars per Replacement Equipment Salesperson.

This ratio measures how effective your salesperson is at generating replacement sales. The benchmark here is $800,000 to $1,000,000 per retail sales person.

$$\text{Sales Per Salesperson} = \frac{\text{Total Replacement Sales}}{\text{Total Number of Salespersons}}$$

There often comes a time when a company's owner can no longer wear all the hats of manager, installer, service tech, and water boy. I am often asked, "Do you think I need a retail salesperson for my existing home replacement business? If so, how do I pay him/her and what kind of volume should I expect?" My response is always the same: "That is a loaded question!" I follow that with asking about your lead generation program and how much you plan to spend. If the answer is "I plan on advertising" but you tell me the amount going to be less than 5% of sales, I warn you not to expect much!

If you told me when I was a student how big a part math plays in life, and especially in business, I would have said "no way." But in reality, it all comes down to math. Here's what I mean.

The industry average closing rate for an HVAC retail salesperson is 28%, or about 1 in 4 estimates. The good ones average between 50–60%. So let's meet in the middle and use a 40% closing average. A retail salesperson needs to sell a minimum of $800,000 to pay for themselves, and ideally should sell around $1,250,000 as a top performer.

Now you can see why I ask how you're going to feed the machine if you hire a retail salesperson. First, to reach $1,250,000 per year, you must sell 192 jobs, or roughly four jobs per week. Let's look at that math:

$1,250,000 / $6,580 avg. system sale = 192 jobs. 192 jobs divided by 50 weeks (assuming two weeks' vacation) is 3.84, or 4 jobs per week.

The investment in lead generation to support this sales effort can be figured by the number of jobs divided by the closing rate, which is then multiplied by the average cost per lead. For the HVAC industry, the average cost per closed lead resulting in a sale ranges from $350 to $500 depending on the advertising media selected.

In our example, we would need to generate 480 leads at a 40% closing rate to sell 192 jobs. Since the average cost per closed lead is around $350, our advertising investment would be $67,200 to support the efforts of a retail salesperson. The formula looks like this:

jobs / closing ratio = closed leads x $ per lead = advertising investment

Sample: 480 leads x 0.40 close rate = 192 jobs sold x $350 average cost per lead closed = $67,200

I know when you first read this, you probably thought, "No way does it cost that much!" But remember—in the beginning, I said you should be spending at least 3–5% of sales on advertising to keep leads coming into the business. That supports the operational costs to do business and lets you make a profit. In this example, $67,200 is 5.3% of sales ($67,200 / $1,250,000 = 5.3%).

To prove the math, let's apply these figures to a real-life advertising experience. One of the most popular forms of traditional advertising used today is direct mail postcards. This is because direct mail solicits a direct response for a call to action based on an offer. But it's only effective if the target audience is in the "buy or need" mode for your product within a short window of time around when they received your direct mail piece. Because the target audience in the buy or need phase is so small, direct mail has an effective return rate average of about 1.5%.

The average cost for a 5 x 7 postcard is $0.47, which includes postage. At this rate, you could send 31,584 postcards to potential customers for $67,200. With an effective average rate of return of 1.5%, this investment would yield roughly 474 leads. With a 40% closing rate, you would sell 190 jobs.

This cost per closed lead ratio applies for most forms of traditional advertising. By this, I mean traditional media like television. radio, and direct mail. The advent of the internet has reduced the cost per closed lead significantly, but internet strategies aren't easy to manage.

Internet marketing involves keyword searches and navigating Google's constantly changing rules about how key phrases are searched or how fresh the content on your website is. Running an effective website and keyword search marketing program requires time and computer skill sets that many contractors do not have. Thus, these services must be subbed out or hired for, which can run up costs for these programs.

The message here as it pertains to advertising and its effect on a retail salesperson's performance is that without an investment of 3–5% of sales, it is harder and takes longer to grow your business, and it's difficult to support or justify a retail salesperson.

Overhead-Related Manpower Ratios

It is obvious that controlling labor costs plays an important role in contractor profitability. Because direct labor is a variable cost in relationship to the workload, it's tricky to know how much fixed labor (i.e., office employees) is needed inside to support the activities generated by the outside production employees. Obviously, the more outside (revenue-producing) employees you have compared to the number of office (overhead-generating) employees, the better. There is a risk, however, of having too many outside employees in relation to the number of office employees. Specifically, you may see inside team burnout or high turnover if the workload inside is greater than the number of people needed to perform it adequately and in a timely manner. If people are overworked, they feel exhausted and underappreciated—and often quit. Hiring and training new employees is costly. Falling behind on sending invoices to customers in a timely manner wreaks havoc on cash flow and the ability to pay vendors without incurring financial charges. These things all affect profitability. The staffing ratio helps contractors know if they have the right amount of office help to support the revenue-producing activities of the outside team.

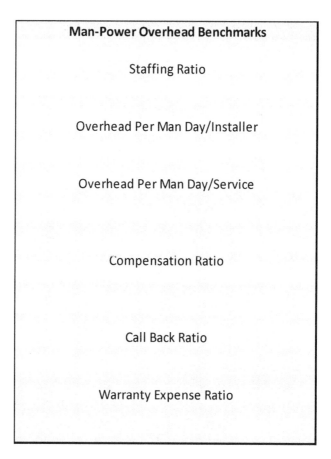

Office to Field Staffing Ratio

Your office to field staffing ratio is of the most important metrics, because it balances how many inside staff (overhead producers) are needed to support the outside staff (revenue producers). A higher staffing ratio demonstrates management's ability to control overhead labor costs. The benchmark for this ratio is 3:1 or higher. This means three outside staff to one inside staff. At a minimum, the ratio should be at least 2:1. Ideally, having three to four people outside to one inside is the most efficient ratio. Anything higher starts to tax the inside team.

$$\text{Staffing Ratio} \quad \frac{\text{Total Field Staff}}{\text{Total Office Staff}}$$

Installer and Technician Overhead Recovery Rates

As I mentioned before, your installation department and service department have different overhead dollars and percentages assigned to them. Knowing how much overhead per day must be recovered by revenue dollars produced by an installer or service technician can assist contractors in establishing a pricing methodology centered on profitability. You can only calculate these ratios if you have a departmentalized income statement that breaks out installation sales and overhead from service sales and overhead. There is a huge advantage to cost control and job pricing when you have departmentalized income statements.

Overhead Dollars per Installer to be Recovered Each Day

This ratio determines the overhead dollars per day by installer that the company must cover with gross margin dollars to break even or make a profit. The industry average is $275 to $350 per day. The reason for the wide spread in the average is the difference between top-performing contractors who understand service pricing models and those who do not. The average installation department overhead percentage for top-performing contractors is around 32–35%. Lower performing contractors who do not have sound installation pricing models will have overhead percentages between 40–45%. This difference is huge when you look at installation department net profits. Top performers are earning 12–15% net profits as an average, whereas lower performing contractors are earning 3–5% net profits. A contractor making a 15% net profit on installation revenues is making 66% more net profit than the contractor who is netting 5%.

$$\text{Installation Overhead Per Day/Installer} \quad \frac{\text{Total Installation Dept. Overhead}}{\text{Total Number of Installer Man Days}}$$

To find the overhead per installer per day, you need to take the total overhead of the installation department and divide it by the number of installation man days per year. This number is typically 240-man days per year per installer (365 days − 104 weekend days −10 vacation days − 6 holidays − 5 sick days = 240). This calculation assumes your company can sell enough jobs to keep your installers working all 240 days. If not, this calculation should factor in the anticipated non-production days and a new overhead per day should be figured into your company's budget.

For example, let's say a top-performing contractor has installation revenues of $850,000 and an installation department overhead of 35%, or $297,000, with four installers. His man days for installers would be 4 x 240 man days = 960 man days.

To figure the overhead per man day, you would take the installation department overhead of $297,000 / 960 = $310 overhead dollars per man day per installer. In this example, you would need to recover $1,240 installation department overhead dollars per day for this company ($310 overhead dollars per man day x 4 installers).

Breakeven per Installer Overhead per Day

Once you have determined your installer overhead per day, you can find out what your daily breakeven revenue must be per installer to cover the overhead per day per installer. This is determined by dividing your daily overhead by the gross margin percentage generated by your installation department.

Keeping with our example, the overhead per day per installer is $310; let's say the gross margin percentage of the installation department for this example is 40% (0.40). The math would be $310 / 0.40 = $775. At a 40% gross margin, you must generate $775 of revenue per day per installer just to break even; for our group of four installers, it would be 4 x $775 = $3,100.

Another way to look at this company would be to say it must sell a furnace and a coil or an air conditioner and a coil every day, Monday through Friday, just to break even. I understand this is not how the workload comes in, but if we were to spread the workload over every day the business is open, this is what it would look like. As it turns out, seasonality makes the HVAC business feast or famine, which adds another layer of complexity into managing the business. That's all the more reason why profit maximization is necessary to provide cash flow through the ups and downs of the seasons!

Overhead Dollars per Service Technician to be Recovered Each Day

This ratio determines the overhead dollars per day by service technician that the company must recover in gross margin dollars. The industry average is $200 to $325 per day. The reason for the wide spread in the average, yet again, is the difference between top-performing contractors who understand service pricing models and those who do not. The average service overhead percentage for top-performing contractors is around 45%. Lower performing contractors who do not have sound service pricing models will have overhead percentages between 55–60%. This difference is huge when you look at service department net profits. Top performers are earning 15–20% net profits as an average where lower performing contractors are earning 5–7% net profits. A contractor making a 15% net profit in service revenues is making 53% more net profit than the contractor who is netting 7%.

$$\text{Service Overhead Per Day/Technician} = \frac{\text{Total Service Dept. Overhead}}{\text{Total Number of Technician Man Days}}$$

To find the overhead per technician per day, you need to take the total overhead of the service department and divide it by the number of service man days per year. As we saw previously, this number is typically 240 man days per year per installer (365 days – 104 weekend days –10 vacation days – 6 holidays – 5 sick days = 240).

To demonstrate this ratio, let's say a top-performing contractor has service revenues of $275,000 and a service department overhead of 48%, or $132,000, with two service technicians. The man days for service technicians would be 2 x 240 man days, or 480 man days. To figure the overhead per man day, you would take the service department overhead of $132,000 and divide by the man days, figuring $132,000 / 480 = $275 overhead dollars per man day per service technician. In this example, you would need to recover $550 service department overhead dollars per day for this company ($275 overhead dollars per man day x 2 service technicians).

Breakeven per Service Technician Overhead per Day

Once you have determined what your service technician overhead per day, is you can find out what your daily breakeven revenue must be by service technician to cover the overhead per day per technician.

This is determined by dividing your daily overhead by the gross margin percentage generated by your service department.

Keeping with our example, the overhead per day per service technician is $275. Now let's say the gross margin percentage of the service department for this example is 60% (0.60). The math would be $275 / 0.60 = $458. At a 60% gross margin, you must generate $458 of revenue per day per service technician just to break even; for our group of two technicians, it would be 2 x $458 = $916.

If the average service ticket is around $219, we can see that this company would have to run four service calls every day, Monday through Friday, just to break even. Again, I understand this is not how the service calls come in, but if we were to spread the number of service calls over every day the business is open, this is what it would look like. As mentioned, seasonality makes the HVAC business feast or famine, complicating business management and giving yet more reason to maximize your profits so you have adequate cash flow through the seasonal ups and downs.

Managing Overhead Payroll

Compensation Ratio

Managing support labor, known as office and management payroll, is just as important as controlling the labor on the job. This compensation ratio measures how much office payroll the company is paying in support labor, including outside salespeople, as a percentage of the sales dollars being generated by the company. Often, when times get tough, we think the immediate solution is to furlough some of our outside staff because business is slowing down. But we sometimes need to look at support labor as well. Much like the staffing ratio, this benchmark will help identify whether our inside payroll is the culprit and whether it's a better choice to lay off some inside help instead. The benchmark here is less than 22%.

$$\text{Compensation Ratio} = \frac{\text{Overhead Payroll}}{\text{Total Sales}}$$

Technical Performance Measurements

The last area of measurement using manpower ratios measures how efficiently and effectively the work was performed. When we send an installation team to do an install or send a service technician on a repair call, our expectations are that the work will be done correctly. We know this is not always the case. Call-back and warranty work are profit eaters. All contractors experience these service or installation mistakes, but you must understand how much they are costing you and, more importantly, how they happen in order to preserve profit and protect your company's reputation as a quality contractor. Using efficiency ratios that measure call-backs and warranty calls helps keep these costs in check and highlights any deficiencies in service technician and installer skill sets.

It must be noted that these efficiency ratios offer valuable information regarding installer or technician performance, but information is only the beginning. It's even more important that you have a process in place that not only records the occurrence but also includes corrective action steps to limit similar problems from happening again in the future.

Efficiency Ratios

Call-Back Ratio

This ratio measures the skill level of a technician's performance. The benchmark here is that call-back labor expense should be less than 2% of service labor sales. Anything higher than 2% should raise eyebrows about service technician competency—you many need to address training issues or deficiencies. To use the call-back ratio, you must have a system that records service labor used on call-back work.

$$\text{Call Back Ratio} = \frac{\text{Call Back Labor}}{\text{Total Service Sales}}$$

Warranty Expense Ratio

This ratio measures the skill level of an installer's performance. The benchmark here is that warranty labor expense should be less than 3% of installation sales. Anything higher than 3% should have you

Scott Ritchey with Gary Kerns

looking at your installers' skills and whether you need to adjust training. To use the warranty expense ratio, you must have a system that records service or install labor used on warranty work.

$$\text{Warranty Expense Ratio} = \frac{\text{Warranty Labor}}{\text{Installation Sales}}$$

Benchmarking Your Efficiency

This benchmark table for efficiency analysis can be used to measure your company's performance and compare it to the industry benchmarks.

Man-Power Ratio Calculation Formulas		Industry Benchmark	Your Numbers
Sales $ Per Service Truck	Total Service Dept Sales / # of Service Trucks	$150,000 - $200,000	___
Replacement Sales $ Per Installer	Total Replacement Sales / # of Installers	$300,000 - $400,000	___
New Construction Sales $ Per Installer	RNC Sales / # of Installers	$170,000 - $225,000	___
Sales $ Per Employee	Total Company Sales / Total # of Employees	$140,000 - $175,000	___
Sales $ Per Replacement Sales Person	Total Replacement Sales / Total # of Replacement Sales Persons	$800,000 - $1,000,000	___

Man-Power Productivity Ratio Calculation Formulas		Industry Benchmark	Your Numbers
Staffing Ratio	Total # Field Staff / Total # Office Staff	3 or 4 to 1	___
Overhead Per Man Day/Installer	Installation Dept. Overhead / Installer Man Days	$275 - $350 Per Day	___
Overhead Per Man Day/Service	Service Dept. Overhead / Service Tech Man Days	$200 - $325 Per Day	___
Compensation Ratio	Overhead Payroll / Total Sales	≤ 22%	___
Call Back Ratio	Service Call Back Labor / Service Labor Sales	≤ 2%	___
Warranty Expense Ratio	Service/Installation Warranty Labor / Total Installation Sales	≤ 3%	___
Service Sales to Replacement Sales Ratio	Total Service Sales / Total Installation Sales	25% to 35%	___

MAKE MORE MONEY: 12 PROFIT PILLARS FOR HVAC CONTRACTOR SUCCESS

Chapter Summary

If you didn't have four years of business school or an accounting back ground I can understand why benchmarking could be a daunting task and seem like a foreign language. I can also appreciate someone asking, what do I need that for anyway especially if I'm making money? I hear you, but I'd like to ask you a question; are you making enough? Are you managing the business to maximize the amount of money your business can and should make? Benchmarking is the tool that that can clearly show you how financially stable you are and how you measure up to the industry. Knowing every detail about your business, to be exceptionally successful, should be common knowledge, as common as every detail an avid bass fisherman would calculate for a successful day of fishing.

Knowing where to put the lure is the first step to a successful day. The best anglers know all the conditions that affect bass activity and feeding habits. They know when the sunshine's on the water that the bass take cover in grassy areas, under lily pads or boat docks. They understand how water temperature affects fish movement and feeding activity and they know they must match the hatch. Making sure the lure imitates the forage the fish are feeding on.

You may or may not have an interest in fishing success but the broader point is, knowing the details about your business will make you more successful. The solvency ratios show you how financially stable you are to handle any unexpected downturns in your business. Knowing and having the correct amount of working capital is essential to managing and funding the daily operations of the business and telling you how much business you can take on and do in the future.

Return and productivity ratios provide details of how well you are managing your assets and how well your people resources are returning wealth and profitability for your business. Understanding these details will give you the business acumen to be better and more successful than your competition.

The best advice to offer regarding benchmarking is a quote from Notre Dame's legendary football coach, Lou Holtz. "In the successful organization, no detail is too small to escape close attention." Benchmarking provides the details that should get that close attention.

Scott Ritchey with Gary Kerns

Chapter 9
The Service Tech Goldmine

Chapter Overview

In this chapter, you will learn:

- How a good service department breeds replacements
- How quiet market share can grow your business
- The low-price myth
- The labor shortage threat
- Increased sales and profits: The pendulum swing theory

Service Breeds Replacements: The Service Tech Goldmine

HVAC contractors work in an industry based on need and timing. The only time consumers think of their industry or service is when they are experiencing discomfort because their heating and cooling system is not working. As I already mentioned, they do not have the same relationship with contractors that they do with the family doctor, dentist, or grocery store. This makes it hard for HVAC contractors to maintain long-term relationships with their customers. It plays into the adage, "out of sight, out of mind."

To complicate things further, we must recognize that the average consumer may only purchase one to two heating and air conditioning systems in their lifetime—but this is where the big money is for most contractors. If a customer does not already have a relationship with your company, chances are you must compete with two other contractors for the job; worst case, you may not even get an opportunity to bid the job at all!

For reasons like these, to effectively compete in the residential replacement business today, contractors must develop a proactive service department and understand the secrets of quiet market share.

Scott Ritchey with Gary Kerns

Although we mentioned the concept briefly in an earlier chapter, let's dig into it more now.

"Quiet market share" is a term used for sales that are generated by a progressive service department that recognized and closed a sales opportunity before the consumer put that opportunity out for bid. Understanding quiet market share and how to get it can create two additional equipment sales per week per service technician. The average contractor with $1M in sales revenues usually has three service technicians. A company that has trained its service technicians to recognize quiet market share opportunities can add up to six additional equipment sales per week in season and one additional equipment sale out of season.

Think about that for a minute! If the average replacement air conditioner with an evaporator coil is roughly $3,500, then that is an additional $7,000 in revenue per week per service technician. For the contractor with three service technicians, that is $21,000 in quiet market share revenue per week—sales that your competitors did not even know were available because they never went out for bid. If we can agree that the average summer cooling season is 12 weeks long, then by understanding the quiet market share concept you could add an additional $252,000 in replacement equipment sales during the summer months with three technicians.

At this point, your skepticism is probably high and the phrases "I can't believe that," "this can't be true," or "that doesn't work in my market" are bleeding through the pages of this book. The fact is, it *is* true, and hundreds of contractors around the country who know how to deploy the quiet market share program are seeing service truck revenues as high as $1.2 *million*. No, that isn't a typo. One Phoenix-based company I know has a service technician who does that much in revenue per year. I will agree that is a very high number and not the norm, though. So, what *is* the norm? Contractors we've worked with are seeing revenues ranging from $300,000 to $600,000 per year per service truck.

Let's put the quiet market share discussion on hold for as second so I can stress the importance of this concept and how it protects you from the number-one threat to your business. Before I identify that threat, let's rule out what it is *not*. Your number-one threat is not low prices in the marketplace; this was proven through Harvard's Good-Better-Best study on consumer buying behavior in the mid-1990s. When given a choice of three options in a product category, with each option containing its own set of unique features and benefits and starting with a "good" option, a higher priced and more featured "better" option, and a "best" option with an even higher price and more features, the average consumer picked the "better" option 68% of the time. The "best" option was selected 8% of the time, leaving the lowest-priced option selected 24% of the time. As it turns out in the HVAC industry, we are often misguided by poor or inaccurate information when it comes to consumer buying habits.

The Low Price Myth

The first myth we have to overcome is that people are looking for the lowest price. Where does this idea come from? Well, there are usually two sources for this misconception.

The first originates from the consumer when they say, "Do you have any deals going on right now?" or "I am looking for a good deal". This is often misinterpreted by the HVAC trade as the consumer wanting the lowest price. But "lowest price" does not mean deal. Most consumers—a whopping 76%, as the Harvard study showed—are looking for *value* in their purchase. This is what dictates price. If the consumer believes the value of the proposition matches the price that is associated with that offer, they think it is a good deal and will make the purchase decision. So, a good deal has value and should not be mistaken for low price.

This leads me to the second source of the low price myth. As it turns out, I already addressed it earlier in this book—and understanding it requires self-reflection. That's right: you need to look in the mirror. Remember, one-third of HVAC contractors fail each year, while 47% fail by year 4. Why? Because their prices are too low. The fear of being priced competitively leads contractors to price low so they win the job. They are basing this on competitive bids, but the problem is that the competitive bid is too low.

I believe the misperception that you must be the lowest to get the job is the number-one killer of contractor profitability. This, however, is *not* the number-one threat to your business. That's because low prices are self-inflicted and can be fixed by you and a good consultant.

The Labor Shortage Threat

No, the major threat to your business is the dwindling supply of labor entering the HVAC industry. Without service technicians, you do not have a service business—and without a service business, you have a very limited and small replacement business.

Remember: service breeds replacements. When a consumer's air conditioning or heating system is not working, most consumers are not thinking they need to replace the system; they are thinking, "This system needs to be fixed," so they call for service. Pre-planners, or people shopping for a replacement system while their current system is working account for 20–25% of system replacement sales. The key word here is *shoppers*. Since the system is working, they can get as many bids as they want to compare prices and value. This is the pressure cooker environment that makes most contractors feel like they must compete on price, too often caving to a lower price to get the job.

In our experience, if a contractor can create more opportunities with reactionary consumers experiencing a system failure or breakdown, they will sell more equipment with a higher closing rate at higher prices because they have less competition—plus, the consumer's urgency level to get back to being comfortable leads to quicker decision making. That was a mouthful, but it is so true.

From this, the solution sounds like hiring more service technicians. The stark reality is different: the odds of implementing this solution are slim to none. Too many studies show that the number of service technicians is in serious decline and the prospects of a younger generation entering the trade is not enough to fill the void. As it stands now, the HARDI Association (Heating and Refrigeration Distributors International) recently published a study concluding that by the year 2021, there will be 221,000 service technician jobs available with substantially less candidates to fill them. In Kentucky and Indiana today, there are 7,032 service tech jobs open and unfilled. In southern states like North Carolina (7,282), South Carolina (3,695), and Virginia (7,213), the stakes are even higher. Although there are many initiatives being taken by the HVAC industry, technical school programs, and government agencies to attract younger skilled talent to fill these vacancies, the prospects for a near-term solution are bleak.

The answer to this problem reports to work every day. It is the service technicians that you currently employ. *They* are the answer to more revenues, higher net profits, and more equipment replacement sales—and they're far more effective than any marketing campaign you might run. *They* are the goldmine you are searching for...but to mine it to its fullest potential, you must understand how it all comes together.

It all starts by knowing how much gold is in the mine.

When we consult contractors about their service department sales potential and overall service, the first question we ask is, "How large is your market served, and do you know how many service calls per year are available?" This is very important information, as it will help us identify what kind of resources you will need to handle the sales volume from these opportunities. These resources include the number of service technicians and trucks to support the call volume, inside support in dispatching and outbound calling, and advertising support to funnel leads to the service department.

Looking at all this, you might say, "I am going to need a lot of people," and that can become expensive, raising overhead. But as College Game Day analyst and former college head coach Lee Corso would say, "Hold on my friend!" My question is, where are you going to find them? Adding to my point, your problem is not available work. The problem is figuring out how we can maximize your company's potential with the resources you currently have, because skilled technicians are in short supply.

Current data suggests that the average residential service truck generates $175,000 to $200,000 annually. Contractors we work with using our quiet market share system are seeing their service trucks produce annual revenues between $300,000 and $500,000—that's an astounding 2–3.25x difference!

How would you like to improve your service department's productivity 100–300%? Increasing productivity is the only sure way to give your company a fighting chance at growing your service and replacement business in an economy that has a short supply of skilled service technicians. The quiet market share program offers another form of protection for your company against the short supply of technicians. This problem is a real threat to your survival, and one you will want to control. As there are fewer technicians to do the available service work in your market because of retiring senior techs, your competitors will be forced to steal technicians from you by offering higher wages.

The only winner in this battle will be the service techs. In the next five to seven years, we see non-union service tech hourly wages possibly reaching $35 per hour, creating pricing pressures for everyone. With the quiet market share program, you can keep your technicians, control these wage increase concerns, and still reward the service technicians for greater productivity that will produce higher net profits for your company.

Before I explain the quiet market share program in more detail, let's first examine how large the service potential is in your market.

Using the formula we used to determine how many replacement opportunities you have in your market, we can also determine how many emergency service breakdown calls will be available in your market. Let's say the population of the market you serve is 55,000. Taking the average of four people per household would give you 13,750 homes. In our experience, a population of heating and air conditioning units will experience an annual failure rate of 25–28%. The 2010 Decision Analytics study for Honeywell found that 24% of homeowners are under a service agreement contract with an HVAC contractor. Further research showed that homeowners who did not have a service maintenance agreement had an emergency service breakdown rate of 4 in 10 homes, or 40%.

In our population of 13,750, the service number of homes with service agreements and emergency breakdowns would look like this: 13,750 home times 24% with service agreements equals 3,300 homes covered by service agreements. If you deduct the number of homes with service agreements from the total number of homes, you can determine the number of homes that will require emergency service due to system failures. So 13,750 homes minus the 3,300 homes under service agreements equals 10,450 homes that are in the pool of the 40% breakdown population. Doing the math, we discover that

10,450 times 40% leaves us with 4,180 that will need emergency service due to system failure in a 12-month period. To figure the actual breakdown rate for the population, you must take the 4,180 homes and divide by the total home population of 13,750, which yields a 30% failure rate.

There is one more step to figure the total number of breakdowns that will occur in a market. We must remember just because a system is under a service agreement does not mean it will fail. Therefore, if we take the 30% failure rate times the number of homes under service contract, we get an additional 990 homes that will need an emergency service repair in the 12-month period, giving us a total market size of 5,170 homes that will need repair work.

If this were your market and you wanted to figure your service call market share, you would take your number of emergency repair service calls and divide them by that 5,170 figure. For example, if your company ran 350 service calls, then your service call market share would be 7% rounded.

From this data, we can extrapolate some interesting information. We can find the dollar market size for emergency repair service due to system failures, and we can find the dollar market size for service agreements. According to Data Analytics, the average service call is $336. Working it through, $336 (falling within out range of $250 to $350 nationally) times 5,170 homes that need service repairs equals an emergency repair service market size of $1,737,120.

It is our experience that a typical spring and fall service agreement performing simple routine maintenance sells for anywhere between $125 to $175 dollars, If I split the difference and use $150 as my average, then the service agreement market size is $150 times 3,300, equaling $495,000—which gives a residential service market size of $2,232,120.

It is important to note this is residential only. The commercial service business is an entirely different animal. As I mentioned before, you can have a wide range of commercial service agreement types, ranging from routine maintenance like belts and filter replacements to quarterly tune-ups or full-coverage labor and parts replacements. Because of this, it's extremely difficult to determine commercial service market size. Remember: if you are going to engage in this type of work, then you should make sure your commercial truck revenues are performing to the benchmark of $250,000 to $275,000 per truck.

When we work with distributor salespeople, we encourage them to have fun teaching contractors how to extract the gold out of the contactors' service tech goldmine. Why have fun? It is far more rewarding to help unlock the hidden potential of a contractor's business success than to constantly be asking, "Do you have anything for me today?" Our research shows that distributor salespeople who engage with the contractor to help solve issues in their business bring more value to the contractor–distributor business relationship, which means both parties enjoy a greater level of success.

The bottom line here is that relationships built around equipment brand names have more built-in limitations to contractor success than relationships geared around sharing best business practices, providing real competitive advantages for both the contractor and the distributor.

The Service Agreement Goldmine

Let's start mining the gold in the area of service agreements. In many of the companies we have consulted with, we see a huge disconnect in the relationship between the business owner and the service tech in their motivations to grow a sizable service agreement business. This is driven by the competing belief systems of the owner and the technician. Too often, the contractor believes that service techs already make high enough wages, so they offer lower bonus incentives, or spiffs, to service techs for selling a service agreement—usually around $10-$15 per service agreement. Meanwhile, the service tech believes that he or she is not a salesperson. They are paid a good wage to fix things and that is what they do—and all that they do.

What we try to do is bridge the gap by showing the contractor that a larger spiff of $50 for each service agreement sold saves the contractor 700% in the cost of a lead. Remember, not all leads turn into sales. Now, you may be wondering where that 700% savings cost comes from. I mentioned earlier in the HVAC system replacements section that the average cost of a lead is between $350 and $500. If I use the lower number of $350, then we can see that $350 is seven times greater than the $50 spiff—that's 700%.

But savings isn't everything. The very best thing about a service repair call lead turning into an agreement sale is creating a lifetime customer. If treated right with service and value, that customer becomes an ongoing revenue stream for the contractor. How big a revenue stream? That's a great question. Our research shows it can be up to $725 per service agreement.

For this number to work, you need a minimum of 500 service agreements. Let's do the math. First, we know that the average air conditioner has a 15-year life. This tells us that 7% of the service agreement population will need a new air conditioner and coil for an average sale of $3,500. In our 500-service agreement population, that is 35 air conditioners, or $122,000 in revenue. We also know the average gas furnace lasts 20 years, giving us a replacement population of 5%, or 25 gas furnaces at an average price of $3,500 for an additional $87,000 in revenue from our 500-service agreement population. Additionally, we know that roughly 30% of these agreement customers will need repair service at an average of $336, producing another $50,400 in revenue from the group. Lastly, if we charge $150 per service agreement, we add another $75,000. This gives us a total revenue stream of $334,400, or $669 per agreement.

If you use the Decision Analytics study data, the average air conditioner replacement sale is $4,300. This change would make the overall revenue stream $362,900 versus $334,400, making the average service agreement revenue $726 per agreement. National averages tend to be higher, and therefore we say you can make up to $725 per agreement.

These numbers should motivate any contractor to aggressively pursue a robust service agreement business. The Decision Analytics study showed that 24% of consumers have a service agreement, telling us that one in four homeowners see value in this service product. Therefore, service techs do not really have to sell this product—they can merely mention it as another service that your company offers. The product sells itself.

But this is not enough to overcome the service techs' belief that they are not salespeople. Just as the math demonstrated why a contractor should pursue the service agreement business, the same mathematical approach can work for the service tech.

In many interviews with contractors, we are often told the average service tech makes about 20 service calls a week. Figuring that 24% of these homeowners will purchase a service agreement when offered, the tech will sell five service agreements per week. If the contractor agrees with us and pays a $50 spiff for new service agreements, the service tech can earn an additional $250 per week, or $12,500 per year. After taxes at a 28% tax rate, the tech is bringing home an extra $750 per month just for giving 24% of homeowners what they already want! This is a lot of disposable income that can be used on a bigger home for his or her family, a nice new ride, money for childcare, and so on. In addition to the wage increase, the service tech is also guaranteeing an additional 500 hours, or 12.5 weeks, of work for him- or herself, virtually eliminating any downtime during seasonal lows. If this opportunity does not motivate your techs to sell service agreements, then you have the wrong service technicians to help your company grow to the next level without increasing overhead costs.

If they still resist the idea of selling service agreements, then they will also resist the quiet market share program, which is the biggest opportunity in the service tech goldmine. As I explained earlier, there is roughly a 20% chance that a repair service call will turn into a system replacement sale. How the service tech handles this situation is critical to your company's success at capturing quiet market share opportunities.

There are two rules that must be followed before you can effectively use the "Pendulum Theory," a key component of the quiet market share program. We must first recognize that the homeowner who initiates the discussion about a new system on a repair service call is lying to us and merely shopping for a price. I know you cannot believe I just told you not to trust a homeowner who said they were interested in a new system, but you will have to trust me for now.

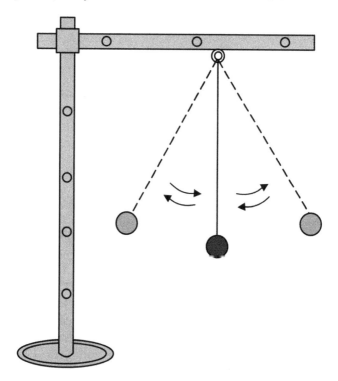

The second rule is that while you want them to be mentioning service agreements, service techs really aren't salespeople. They do not want to be perceived as one and you do not want them to come across as one in a quiet market share opportunity. Once a customer sees your tech as a salesperson, the opportunity is dead. Remember: in the customer's eyes, the service technician is the most trusted employee you have. They are there to solve the customer's problems, not to take advantage of them by selling them something they do not need and the tech does not want to sell.

By now, you are probably saying, "Scott and Gary, you are confusing the hell out of me!"

We know.

So let's take you back to your service tech/owner days when you were just getting started. How often did you hear this phrase from a customer whose system you were working on: "Jim, I was just wondering how much a new system would cost?" Too often, you made the mistake of giving them a ballpark price because you thought they were interested in buying one. You gave them a price and they said, "Thanks, Jim, that sounds pretty good. When I get ready to replace it, you're the one I will call."

How often did that customer call back and purchase?

Not too often, I'll bet!

This is an example where both rules applied. The customer was lying and used your information to shop for a better price, and the service tech/salesperson line was crossed when you gave a price so quickly without any more customer engagement.

We like to call the customer engagement process, "owning the pain by the customer." When the customer owns the pain and concludes on their own that the resolution is a new system, then the technician is not viewed as a salesperson. The customer is in control of the purchase decision and is ready to buy. Anything less and the conversion from service repair to system sell is doomed. This is why we teach contractors and service technicians the "Pendulum Swing" technique.

The Pendulum Swing Theory

The Pendulum Swing Theory is based on Newton's First Law of Motion, which states that a body at rest tends to stay at rest, and a body in motion tends to stay in motion with the same speed, and in the same direction, unless acted upon by an unbalanced force.

Let's create a visual so you can better understand this concept. Start by drawing a circle and number it like a clock.

Now, draw a thick solid line from the 9 o'clock to the 3 o'clock position. On the middle of the line, we will place a pendulum. The pendulum can swing back and forth from 9 to 3 and from 3 to 9.

This represents how your prospect feels about your opportunity. At any given time in the sales process, a prospect can feel positive, negative, or indifferent toward what you are selling.

Below is how your prospect feels toward your offering, relative to the pendulum:

Nine – The prospect is not interested (it's over)
Eight – The prospect is hostile
Seven – The prospect is negative
Six – The prospect is neutral (indifferent)
Five – The prospect is open to looking
Four – The prospect is excited about your offering
Three – The prospect has decided to buy (done deal)

The pendulum will move back and forth. The service tech's job is to keep it swinging until it reaches 3 or 9, because that will be the only time at which a prospect will make a committed decision either way. Remember: the service technician is not selling; they are merely having a conversation with the customer. This conversation will lead the 20% who are really thinking about purchasing but have not owned the pain yet to decide to make a purchase at that moment. It is the service tech's job to handle the conversation in such a way that the customer feels and owns the pain, leading that customer to tell the service technician they want to purchase now.

The shopping customer is the hardest to sell, because it takes a lot of energy to get them moving—their only objective right now is to get a price that they can shop. And without stimuli that will keep the prospect engaged, they will most likely lose momentum and come to rest at 6 o'clock (neutral).

Think about it for a moment. They know they need to consider a new unit and are feeling some pain or they would not even ask the question. The problem is, they just don't own the pain yet and a skilled technician who has mastered the pendulum swing can get them there.

There are a few rules to using this technique successfully. Once you've mastered them, you and your service techs can close more sales and increase your revenue.

The First Rule

The first rule states that technician must supply the energy to keep the prospect moving. You can never depend upon the prospect to do this alone.

The Second Rule

The second rule states that it doesn't matter if a customer is positive or negative on the pendulum; the most important thing is that the customer is *moving*. An indifferent customer will never make a decision (unless it is a decision not to make a decision).

The Third Rule

The third rule states that you should always stay slightly to the left or behind the customer on the pendulum curve. In other words, always remain slightly more negative than your prospect.

The Fourth Rule

The fourth and final rule states that once the customer owns the pain, the service tech should share a story about how they had another customer in the same situation who chose to purchase a system from the technician. This gives the current customer the affirmation that they are making the right decision to replace—but the tech did not ask them to do this.

Let's take a look at why the service tech too often moves right to offering a price when a repair customer baits them by saying, "I am thinking about replacing this old unit—how much does a new one cost?"

When the typical tech hears this, he moves to 3 o'clock on the pendulum and says, "Well, I can help you with that. A standard system usually goes for $3,000."

The problem is, the customer is at 4 o'clock and is not yet sold—and won't be until they reach 3 o'clock (and stay there).

So how does the customer respond to the selling technician? He moves back to 5 o'clock. The customer says, "Well, Johnny, that sounds real good. When I get ready to buy, I sure will be calling you."

Have you ever felt you had one on the line, and then were surprised that they never bought? You would have bet the farm that this person was going to buy. This happens all the time when you use traditional selling methods.

So why does the repair customer react this way? Is it too much pressure or fear? This occurs in the early stages of the selling process, because a certain level of trust and rapport has been established between the customer and technician. When the technician responds too enthusiastically, the customer moves to the protected position of neutral.

In many instances, this is where the technician panics as he sees the potential replacement sale slipping away. He becomes even more enthusiastic and begins to pressure the customer. The typical technician does this by going into a "features and benefits" song and dance. Basically, let's throw the kitchen sink at this customer and hope something sticks. The problem here is that the longer the technician continues down this path, the less likely the customer is to buy.

How does the customer respond to this display of overenthusiasm from the technician? He senses that something is not right and begins moving toward 7 on the pendulum...moving right past neutral into the negative zone. Soon the customer will move to 9 o'clock, set up camp there, and the sale will be lost.

How could this situation be handled to create the desired positive outcome? When the customer is at 4, the technician should move to safe ground—let's say 5, or maybe even 6. By moving to 3 and trying to sell too early, you run the risk of getting between the customer's current location on the pendulum and the point to which we want the prospect to move.

Most technicians get in their own way, and thus become the reason the customer who was shopping did not buy. The shopper got what he wanted in the ballpark price and stepped away from the sale. All the work done to move the prospect to 4 on the pendulum is wiped out in a few seconds.

When a customer is at 5 or 4 and says, "I have been thinking about replacing the unit; about how much does one cost?" the technician's response should be: "Why would you want to do that? I will have you up and running in the next 45 minutes."

This response always catches the customer off-guard and leads them to try again with something like this: "You don't have to give me an exact price—just a ballpark number will be fine." Again, the technician must push the pendulum with a response like, "Really, it's no problem here. I can have you fixed and running soon."

Realizing he is not getting anywhere, the customer will try the recommendation angle: "Johnny, if it were your system, what would you do?" I am sure you have heard that one before! The tech should respond with, "I know you think you want a new system, but can I ask why, seeing I told you I can fix this one?"

Wow, now the customer must describe his pain. "Well, Johnny, this thing has nickel and dimed me a little and I was wondering what you would do." To this, Johnny would respond, "I can appreciate your thinking, Mr. Jones—anytime something starts costing us money, it's a real bummer. But are you sure you've given this enough thought?" This gentle reply will get the customer moving without feeling pressure as he reflects on the pain the older system is causing.

This statement does two things:

1. It offers reassurance to the customer that he is not being sold. He'll feel that the service technician is concerned about the customer's best interests.

2. If there is a hidden objection, it brings it to the surface early so you can handle it. Unstated objections or customers not

owning their pain are major reasons why most technicians lose customers they should have been able to close.

Let's get back to the example.

The prospect moves to 5 and begins sharing their pain points. This is a critical step for the homeowner to come to the realization they need to replace the unit and for the technician to listen so he or she can shape the closing story around another homeowner who had a similar circumstance and chose to purchase a new system.

After the homeowner describes the pain, the technician now would say something like this: "Wow, Mr. Jones, you have put some thought into this. You asked me for a recommendation, but I really don't like doing that because it is your money. But if you're interested, I can share an example where I had a customer, Mr. Nelson, who was in a similar circumstance and what he did."

More often than not, the customer asks for the story and the technician should respond with something like this: "Mr. Nelson was having similar small breakdowns to yours and was facing another repair of $300. He had a budget in mind between $3,500 and $4,500, and after reviewing the options available, he decided to replace his old system. Knowing what his budget was, I was able to come up with a solution that fit and we installed a new system the next day. We could do the same for you."

This is the most critical step in the pendulum swing close. At this point, the service technician should say nothing. I admit, silence can be awkward, but this is where the 20% of customers who want to do something step in and say something like, "You're right, Johnny—I need to quit wasting money on this thing and change it out. What can you do for me?"

Always remember Newton's Law of Motion: A body at rest tends to stay at rest, and a body in motion tends to stay in motion with the same speed and in the same direction unless acted upon by an unbalanced force. And that unbalanced force is the technician.

In quiet market share selling system, the technician effectively resists the bait to give a price upon first request and instead uses reversal questions to get the customer moving along the pendulum. It doesn't matter which direction they move—we just want them moving. Now that the customer is in motion, he will stay in motion.

By using questions, the technician will eventually lead (rather than drag) the customer back to 3 o'clock, where the deal is done...and the customer thinks he got there all by himself. As long as the technician continues to ask questions, sooner or later, the customer will swing up to 3 o'clock. It's the law!

Once your technicians master the pendulum swing, you can use the quiet market share program to its fullest. The combination of a strong service agreement program and the successful implementation of the quiet market share program will take the average service truck from $150,000 to $350,000-plus in one year alone, proving that the service department is the true goldmine of the company.

The companies we have consulted with that have implemented our programs have not only improved net profit efficiency by reaching net profits from 15–22%, they have positioned their companies to protect themselves from the decline in the service tech labor pools. In addition to efficiency, our clients are also able to pay their technicians more, which allows them to easily recruit the best techs in the industry. A typical service tech in our program has an income in the $80,000 to $100,000 range and the best part is, the company they work for still sees double-digit net profits.

We often get asked if this system replaces the need for a retail salesperson. In most cases, the answer is no, because the retail salesperson serves a different type of customer. The retail salesperson better serves the proactive customer. This customer is not in a demand service situation. They usually know they need a new air conditioning system, either by the age of the system or because past repairs have taken a mental and pocketbook toll already. These shoppers, in most cases, have been doing research and have a different set of questions for the retail salesperson, typically pertaining to system comparisons and attributes. This type of customer would take up too much of a service technician's time. Additionally, it requires the stronger selling skill set of a dedicated retail salesperson to influence and close this sale. All the things most technicians hate about the sales process!

Instead, the service tech goldmine is the reactionary customer in a demand service situation where their system is broken. The emotional level of this customer is much higher—they want to get something done quickly to remedy their situation. They'll either repair it or, with the right pendulum technique, replace the system. It's a much better solution for the customer, and a more profitable solution for you under the right circumstances.

Chapter Summary

Understanding the pendulum swing sales process and how to implement it is the single most important component of success in the quiet market share arena. Though it *is* a sales process, the technician should never come across as a salesperson. Nobody likes to be sold, and in general, most technicians do not like to sell. They are great at recommending and offering their knowledge—that's why the pendulum swing process works so well for most technicians.

Remember, in most situations where the process gets started, consumers are trying to get free information from your tech. Usually this free information involves a solution recommendation and a price that they can use against you by taking your information to a competitor. Your technicians must realize that once you're in front of a customer, you own that customer and they need to do everything to ensure that customer stays with you. A big part of that is not giving away free information that can be used against you by your competitors.

Mastering the pendulum swing process will produce a service call to installation sale conversion rate around 15–20%. We have seen rates as high as 23%. We have also seen average replacement sales tickets climb $1,300 to $1,800 higher when the technician sells the job instead of a retail salesperson. We don't recommend replacing your retail salesperson with this process, but you may find that if you currently have multiple retail salespeople, you may not need as many after implementing this technique.

By making this highly effective change to your strategy, you can maximize your company's potential while minimizing risks.

Chapter 10
Technician Training and Compensation Ideas

Chapter Overview

In this chapter, you will learn:
- The need for good technicians
- How compensation is the key to retention
- The labor shortage threat and its effects
- The importance of key benefits like insurance and 401(k)s

The Need for Skilled Technicians and How We Train Them

We've talked extensively about the many benefits of offering well-rounded services. Now it's time to talk about the technicians who make these services happen.

You simply cannot run a good HVAC company without having good service technicians to do the work. Unfortunately, as mentioned, qualified HVAC technicians are becoming increasingly hard to find, and this trend is set to become even more pronounced in the future. It's estimated that the HVAC industry will need more than 220,000 new qualified workers in the next seven years--but HVAC work is not exactly one of the most popular career choices for young men and women, even though the pay is above average and it doesn't take a four-year university degree to learn the skills needed to be successful in the field.

The upcoming retirement of many baby boomers will also worsen this problem. Current stats show that up to half of today's HVAC workers will retire in the next decade. Those who are set to take their place are young and don't have nearly as much experience as the HVAC technicians who are heading into their golden years.

What does this mean for your business? You need to pay attention to your employees if you want to have a successful company that stands the tests of time! You cannot afford to pay your employees less than the industry average of $49,000 a year plus overtime and benefits. You also need to go out of your way to make your company a pleasant place to work. Regular company events, bonuses, and a pleasant work atmosphere will help you keep experienced technicians so that you can continue to take on jobs without worrying about who will be doing the work.

Tech Compensation: The Key to Retention and Profits

We've touched on the importance of adequate compensation for your HVAC techs throughout the book, but now it's time to have an in-depth look at tech compensation. This is no doubt the most highly debated topic when we talk to contractors about recruiting and retaining quality technicians and installers. Business owners have tons of questions for us: How much should I pay someone who is green to the industry? What is a person worth who has five years, 10 years, or more experience? What are my competitors paying their help?

When Gary and I get into these discussions, we first ask our clients to think outside the box. Your competitors are not the only businesses competing for this talent pool. When you look at it this way, you can clearly see that the entry-level pay for green employees is not dictated by what they do or do not know. Rather, it's based on what *other* industries are willing to start them at. And while you look at that, consider what type of work they'll be doing and the physical demands of the job they will be performing at those companies compared to yours.

Companies like Amazon, Ford, and UPS are hiring inexperienced people for $12 to $14 per hour. As of the writing of this book, the unemployment rate was 3.9%. Anything under 5% is considered full employment. The economy is strong, consumer spending is up, and inflation is low. These conditions are great for business, but place incredible strain on your ability to recruit quality help at a reasonable cost. For this reason, the HVAC industry must move to more creative and competitive compensation programs to attract the right kind of talent.

Without adequate tech compensation, your business is guaranteed to become one of the 90% of HVAC companies that fail within 10 years of starting up. While knowing how to manage business finances and accurately calculate overhead costs and other factors plays a key role in the success of every HVAC business, you simply cannot have a successful HVAC business without good technicians and installers to do the work for you. There is just no way to get around it. Furthermore, as we discussed above, there is a shortage of qualified HVAC workers that is only going to get worse in the coming years. Supply is low and demand will grow as an increasing number of homeowners require HVAC services and a healthy housing market demands the installation of new HVAC units on an ongoing basis. Commercial services and installation may not be growing as rapidly as residential work, but rest assured that there will be plenty of commercial HVAC work to keep you busy in a strong economy.

So, what exactly does "adequate compensation" look like? Let's start by talking about salary and then move on to other forms of compensation.

Technician Wages

How much should you pay your HVAC technicians? Truth be told, that depends on a number of factors. The federal Bureau of Labor Statistics makes it easy to see the market cap for HVAC worker wages by state. This is important because some states have higher wage averages than others and you need to meet your state's "wage market cap" if you don't want your workers to leave your company in favor of a competitor's.

If you want an up-to-date idea of what HVAC technicians earn in your state, there are several websites you can review to get salary and wage data by the type of job. Take a look at the Bureau of Labor Statistics website, Salary.com, or Glassdoor.com, each of which outlines these stats in detail. One thing you will probably notice right away is that there is a very large difference in wage expectations depending on where you live. In the southeastern United States, wages for HVAC workers range between $23,660 and $42,880 a year. On the other hand, midwestern cities like Louisville, Lexington, and Indianapolis range between $42,000 to $67,000. The average wage range for HVAC workers in western U.S. states such as California, Nevada and Washington is $66,710 to $75,230. In case you're wondering how companies can afford to pay such high wages, bear in mind that the market cap for service, maintenance, and installation work is far higher in states with high wages than it is in states with low wages. As one Californian contractor accurately put it, HVAC installation work probably costs at least $1,000 more in California than it does in just about any other state.

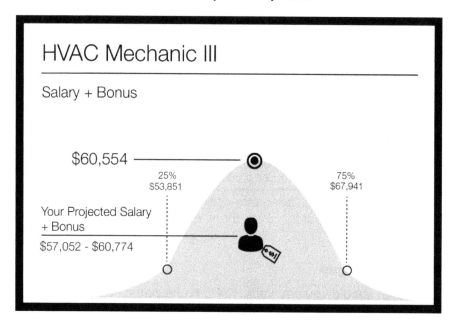

The jobs you expect your HVAC technicians to handle will also determine how much they should be paid every year. As the Bureau of Labor Statistics website points out, the average annual salary for an HVAC technician who handles residential repair and maintenance is more than $3,700 lower than the average annual salary for an HVAC worker that handles commercial HVAC installation and repair. HVAC technicians who know how to install specialized HVAC systems in hospitals command a far higher annual salary than those who install units in a regular commercial or residential building.

Take average HVAC technician salaries in your area into account when you are deciding what to pay your existing workers or how much to offer new HVAC technicians you are hoping to hire. Your workers are not only your most important asset but also the one that is hardest to replace.

Offering Performance-Based Pay: How Does It Work?

Performance-based pay is a system used by many successful HVAC contractors. As the name implies, this system enables you to pay good workers more than you pay average or subpar workers. It's ideal because it not only rewards your employees for doing good work but also gives them an incentive to do even better work in the future. If your HVAC technicians show up on time for their jobs, do good work, have a low call-back rate, and don't generate customer complaints, they should meet quality and efficiency standards to earn extra money.

Furthermore, performance-based pay is not simply a great option for your HVAC technicians; it can also be used for any salespeople that you decide to hire. If your salespeople generate more sales by selling options like indoor air quality or service agreements on each sale, reward them with a spiff program to increase their salary. Those who generate few or no ancillary sales won't earn the extra money. It's as simple as that.

There are many ways to offer a performance-based payment system, so consider the options to see which method works best for you. Perhaps the best way to calculate performance-based pay for your HVAC technicians is to offer an extra sum of money through a commission program for service calls that turn into equipment sales. Companies that use our quiet market share program pay commissions of up to 6% on sold systems depending on the gross margin percentage generated from the sale. As we mentioned before, we are strong advocates for paying service technicians $50 for every service agreement they sell and $25 for every renewal. These methods have earned technicians we have worked with an additional $12,000 to $20,000 per year, providing annual technician wages in the $70,000 to $80,000 range. With wages like these, you can expect your technicians to stay with you and find others who want to work for you, enabling you to grow your company.

Naturally, this system can be a bit of a headache for you, as you will be paying your employees a different amount each week. However, it's not hard to calculate employee wages if you use the formulas we talked about previously to determine the percentage of gross margin needed to cover labor costs.

There are other alternatives to performance pay. You can simply pay a predetermined amount per job, better known as piece pay. There are several complications to this method, however, and if done incorrectly, you might be penalized come tax time.

Therefore, the key to making performance-based pay work is to know the federal and state labor law requirements for workers at your company. As your workers are hourly workers rather than part-time contractors, you need to pay them at least the minimum wage, pay them extra for overtime, and ensure that each worker has a time card that calculates the number of hours worked in a single day. Performance-based pay does not take the place of overtime; instead, overtime payment is added to a worker's weekly pay.

If you aren't sure how to meet overtime payment requirements, follow the Department of Labor's helpful formula for calculating payments:

First, you'll need to determine how much compensation your worker is earning for the pay period. Once you have done this, divide this compensation by the number of hours worked. You will then see how much you are paying your worker per hour. If your employee works overtime, you need to pay the hourly rate plus an additional half.

Here's an example of what this look like in real terms:

Tom handled 30 service calls in a single week. He met company standards of excellence for each call, so he earned $40 per service call. Multiply this by the amount of jobs Tom did (30 in this case) and you will see that Tom earned $1,200 in the past week.

Here's another scenario: Tom put in an extra five hours of work last week because a client wanted some work done on Saturday. This work came from a single job, so Tom earned an extra $40 for doing this job well. His total pay for the week then came to $1,240.

To determine the amount of overtime, calculate Tom's total wages for the week and divide it by the number of hours he worked. If you divide $1,240 by 45, you will see that Tom earned about $27.50 per hour. You then need to multiply the hourly salary Tom earned by working overtime by the number of hours he worked; in this case, this means multiplying $27.50 by 5. Divide your answer by 2 and you will see how much you need to pay him in overtime pay. Add this to Tom's total salary for the week.

It may seem that performance-based pay is taking a big cut out of your net profits. But consider the fact that it is also enabling you to save money. Companies that offer performance-based pay often find that overtime work decreases as HVAC technician have an incentive to do jobs efficiently because their income is based on number of service calls rather than solely an hourly wage. Furthermore, your call-back rate will decrease as your HVAC workers are eager to keep their performance-based pay rather than have to forfeit it due to on-the-job mistakes. It's also worth noting that if each worker on your team uses their time as efficiently as possible, it could also help you reduce the total number of workers you need to hire.

Furthermore, performance-based pay keeps your workers happy. You don't have to worry about offering salary increases or even extra bonuses every year because your employees are already earning the money they want through your pay system. It's great not only for long-term workers but also new workers who come to your company because they don't have to wait until they have a few years of work experience under their belt to earn the money they need to have a good standard of living.

At the same time, performance-based pay also penalizes workers who don't do a good job. You no longer have to berate, nag, or remind workers who make on-the-job mistakes, mishandle customers, or fail to show up for work on time. Simply rescind their performance-based pay and that's that. Your HVAC technicians will either get the point and improve their performance or find somewhere else to work.

Don't Forget Pendulum Swing Commissions

As we touched on in an earlier chapter, allowing technicians to earn a commission on equipment sales is another way to keep them highly compensated. One particularly important benchmark to keep tabs on is the average number of service jobs you are doing each week and month. The average technician should run four to five calls per day, or 16 to 20 calls per week. As previously mentioned, the revenue dollars per service truck will help you manage the number of service technicians you need. Remember, the industry average is around $175,000 to $200,000 per truck, but companies using the pendulum swing technique are making two to three times that. Seasonality *does* influence the average number of service calls, with peak times demanding more calls and valleys offering fewer calls. Keep in mind that it is important that your service technician compensation program rewards your techs via selling spiffs or commissions so they can make money, especially in the peak times, so that you can send them home in the slower times. This keeps their annual income in that $50,000 to $75,000 range, which will keep them employed with you and still allow you to manage your cash flow through the peak and valley times.

Other Forms of Compensation Important to Recruitment and Retention

Regular pay isn't the only form of compensation that matters to workers. You can offer additional incentives to attract the best workers to your company and, by doing so, improve your customer satisfaction and your bottom line.

Health Insurance

As we mentioned earlier in the book, health insurance is currently the most sought-after benefit in the HVAC industry. Health care is expensive, especially for technicians who are married and have kids. Offering health insurance is a great way to keep your HVAC technicians working for you long-term. You may even be able to offer a lower salary than your competitors if you are willing to add health care to the package. HVAC technicians know that health care costs money and many will be willing to work for less hourly pay than their counterparts if they can get this all-important benefit.

The type of health insurance you offer will naturally depend on your net profits, how long your company has been in business, and the type of deal you can negotiate with an insurance agency. While successful HVAC contractors across the country agree that providing health insurance is a must to retain the best HVAC technicians and installers, they also provide interesting insight into what types of coverage they offer. When asked, one contractor noted that he offered plans that paid for up to 70% of an employee's medical costs. He said that he would like to offer plans to his employees that covered 100% of medical costs but simply did not have the money to afford it. Another company owner said that he pays 100% of all employee medical costs; however, his plans did not include either dental coverage or eye checkups. Yet another plan offered by a successful HVAC business owner in a different city included full medical and dental coverage but no coverage for eye exams.

Unfortunately, the cost of providing health coverage to employees is growing every year and the political uncertainty surrounding the Affordable Care Act will probably only make things worse. If you aren't sure about legal requirements for providing certain types of health insurance to your workers, I would suggest talking to a knowledgeable attorney. If you are concerned about keeping costs low while offering great coverage, talk to as many insurers in your area as possible. However, bear in mind that once you start offering a certain type of coverage, you can't simply take it away from your employees without creating employee discontent or even losing good people. You may be able to offset a type of coverage you are taking away with another, better type of coverage for your employees and their families.

However, I would suggest you avoid doing so without talking to your employees first. You may not be able to keep everyone happy, but your employees should see that you are doing all you can to offer them the best possible health insurance that you can afford.

Remember, health insurance isn't merely about getting good employees and making them want to continue working with you. It's also about keeping your employees healthy so they can continue to work with you. HVAC installation, maintenance, and service are all active jobs and your employees need to be in good shape to handle the workload. Health insurance helps your employees stay healthy; failure to offer it results in unexpected downtime from illnesses and conditions that could have been avoided with timely medical intervention.

401(k) Plans

Matching contributions to an employee's 401(k) plan is one of the best benefits you can offer because it doesn't cost you anything. All contributions you make to your employees' accounts are fully tax deductible, essentially taking some revenue that you would have paid in taxes and giving it to your employees instead. You really can't find a better way to compensate your employees for their hard work without dipping into your net profits than to make matching contributions to their 401(k) plans.

Furthermore, matching 401(k) contributions benefit your employees even more than giving them an annual bonus. The extra money you give your employees will be taxed by the federal government. The average worker faces a tax burden of up to 22.4%, which means that 22.4% of your employee's income is spent on taxes. If you give your worker a bonus of $1,000, that worker only really gets $776 when you factor in both his or her taxes and your own taxes.

On the other hand, a contribution to a retirement account is tax deferred. Your employee does not have to pay taxes on this extra income until he or she retires. At this time, the employee is not usually earning much, if any, extra money; this means that his or her tax burden is far lower than it would have been when your employee was working for you full-time.

Scott Ritchey with Gary Kerns

Understanding How It All Works

As the name implies, "matching contributions" means that you are contributing the same amount that your employee is putting into such a plan. If, for instance, your employee decides to put $10,000 in his or her 401(k) account, you add an extra $10,000. This is a 100% matching contribution; however, there are lower levels of contribution as well. Many HVAC companies offer a 100% match on the first 3% of income its employees contribute; additionally, there is an equally graded vesting period so that employees do not immediately benefit from the money that you put into their account. This whole financial concept is complicated, but suffice it to say that you can set up your contributions to an employee's 401(k) account in such a way that your employee will not gain the benefits of these contributions unless they continue to work for you for a pre-specified time period, usually up to five years.

Be aware that there are a plethora of laws in place that dictate what type of 401(k) account you can provide compensation to, how much compensation you can offer certain employees, and who you need to offer compensation to. You cannot contribute to some employees' 401(k) plans and not others. It's a complex subject, but an accountant can help you sort it all out pretty easily.

Furthermore, the federal government even offers compensation to cover the costs of setting up 401(k) contributions and teaching your employees about the benefit of saving for retirement. If your company has fewer than 100 employees, has at least one plan participant who isn't highly compensated, and you paid your employees at least $5,000 in the previous year, you may be eligible for the Credit for Small Employer Pension Plan Startup Costs. You can receive a credit for up to 50% of all your startup costs and claim up to $500 every year for three years.

Life Insurance

Life insurance is a benefit often overlooked by small businesses—especially ones that offer limited benefits like many HVAC contractors. This benefit, however, can be a low-cost way to retain employees with families. Term life insurance will benefit your employees' families should something happen to them. The cost of term life insurance, depending on the age and medical condition of your employees, can be as low as $10 per month. Companies that offer term life policies usually offer a minimum of $25,000, going up to two times annual earnings. These amounts easily take pressure off the family for funeral and burial costs while also providing a few months of income until the family can sort things out after a tragic event like a death.

There are several common policies to pick from. Group Accidental Death and Dismemberment (also known as AD&D coverage) pays benefits to the employee's beneficiary should the employee die or have an accident that causes him or her to lose the use of one or more limbs. Business travel accident insurance only covers an employee if he or she dies while traveling, and you probably don't need to offer it unless you accept out-of-town jobs on a very regular basis. Split-dollar life insurance, as the name implies, is paid for by both an employee and his or her employer. If the employee dies, the employee's beneficiary receives the funds from the insurance plan, while the premiums paid into the plan by the employer are returned to the employer. This type of insurance is usually only offered to key employees. All these life insurance plans can have riders (e.g., additional features and benefits) added to them on request. Naturally, riders change the cost of the policy but may be well worth the additional price.

Life insurance has similar business advantages to a 401(k) plan in that you don't have to shell out all the money you need for life insurance costs from your net profits. Using group term life insurance policies that the IRS considers non-discriminatory can gain you a tax break. Furthermore, buying the right type of policy will prevent life insurance from being counted as part of your employees' taxable gross income.

Life insurance policies are much easier to manage than health insurance policies because you won't be filing claims on a regular basis. However, you do need to be aware of what you will have to do if an employee dies or decides to leave your company. Talk to your insurer about policies and rules before you sign up for coverage. You will also want to talk to your employees about their own responsibilities should they accept life insurance from your company. Your employees are responsible for naming the beneficiary of their policy and changing the name of the beneficiary should they so desire. Your employee also needs to tell the beneficiary about their life insurance coverage, as the beneficiary will need to call you in the event of an employee's demise so you can fill out the necessary paperwork for them to receive the policy benefits.

One very important factor to consider when purchasing a group life insurance policy is that some insurers will ask employees to fill out a medical questionnaire. This is often done to determine the cost of the policy. If you opt to work with an insurer that asks employees for medical information, be sure to keep the information private between the employee and the insurance company. Provide your employees with envelopes so they can put the completed questionnaire inside and mail the information directly to the insurance company. If you accidently see an employee's medical information and then opt to fire him or her for some reason, the employee can sue you, claiming that he or she is being fired for a medical condition that you knew about.

Tuition Assistance

Tuition assistance is becoming far more common than it was in times past. In fact, it is fast becoming one of the most sought-out benefits around, as the cost of a college education soars past what many people can afford to pay.

There are several options to consider when deciding what type of tuition assistance to offer your employees. Large companies often team up with certain educational institutions; they obtain large discounts from certain colleges and universities and then offer to pay tuition if their employees sign up for courses at one of those schools. Other companies require that employees first sign up for federal aid before accessing tuition assistance. Other rules include requiring employees to work with the company for a certain number of years after obtaining a degree, requiring employees to get a certain minimum grade to receive tuition assistance, and setting restrictions on what type of degree can quality for tuition assistance.

You will probably want to limit your tuition assistance to helping employees study a field directly related to their HVAC work. However, you may also want to consider including managerial degrees in the list of programs you are willing to sponsor. Promoting your own workers once they show they can capably manage the workload will save you the time and hassle of having to find a manager who doesn't know anything about your company and then training him or her to handle management duties for your HVAC contracting firm.

Tuition assistance is yet another employee benefit that can be written off at tax time if you follow IRS stipulations. I would strongly advise you to sit down with an accountant and get professional help filing your taxes so you don't lose net income providing benefits that could be deducted from your tax payments.

If you do wind up having to cover some of the tuition assistance you offer out of pocket, take heart. A recent survey by Cigna found that companies that offer tuition assistance to employees not only get the money back but also make an additional $1.29 thanks to the fact that tuition assistance dramatically reduces employee turnover and thus eliminates the costs involved in finding and training new employees.

Paid Vacation Time

The United States is one of the few industrialized countries in the world that does not have laws in place guaranteeing employees paid vacation time. Even so, this doesn't mean that you shouldn't consider offering your employees this time off. Bear in mind that HVAC work is very tiring. It involves a lot of physical activity and your employees will sometimes need a break to sit back, relax, connect with friends and family members, or simply enjoy life on their own.

How many vacation days you offer is up to you. Have a look at what other HVAC companies in your area are offering and provide benefits that match those provided by other good companies. At the same time, you may want to consider a flexible, nontraditional plan that does not separate sick days from vacation days. If you offer generous time off, make sure your employees still need to call and get managerial approval before taking an extended break. Having separate sick days and minimal vacation days encourages your employees to call in sick every time they need or want a day off. This can wreak havoc with your HVAC servicing schedule if you don't have the personnel on hand to attend to existing commitments.

Employee Discounts

Your employees need working HVAC systems just as much as your clients do. Offering them discounts on HVAC parts, equipment, and installation is a great benefit for them that doesn't cost you much to provide. Rest assured that the technicians who work for you will appreciate getting a large discount if they need to replace an HVAC part or have a new unit installed at home.

Employee discounts often pay for themselves in more ways than one. Your employees can become some of your best salespeople, as they benefit from your company's products and services and go on to tell others about them. In other instances, your employees may even purchase parts and services for close friend and family members who in turn become lifelong customers of your company thanks to the fact that they can see the high quality of your goods and services for themselves.

Safety

Safety is not a benefit in the sense of being an optional extra that you decide to offer, but your employees' safety is something you will need to factor in before seeking out commercial or residential HVAC installation jobs. While these jobs do pay well, there are also costs associated with keeping your workers safe as they do their jobs.

Review OSHA's basic guidelines regarding the handling of hazardous equipment, respiratory protection, confined space entry, electrical standards, fall arrest systems, and machine guarding standards before bidding on construction jobs if you want to keep your workers out of harm's way. At the same time, don't think that safety guidelines only apply to HVAC contractors who work at construction sites. Some of the most common types of accidents in the HVAC industry can occur no matter where you work or what type of work you do. Such accidents include electrocution, falls, and burns caused by soldering equipment. Perhaps the biggest danger is fatigue, as overworked HVAC technicians may fall asleep at the wheel or fail to notice that parts and wires have not been properly connected.

Failure to follow safety guidelines can be deadly. Allow me to remind you of a point I made earlier in the book: HVAC work is the 10th most dangerous jobs in the United States. Construction work comes in eighth place on the list. Because it's even more dangerous than HVAC installation, HVAC workers at a construction site are often required to follow stricter rules than those who primarily provide residential HVAC installation, repair, and maintenance.

Not only can failure to follow safety guidelines result in serious injury or death to your workers, it will also likely result in serious legal problems for you and your company. All it takes is one lawsuit by the family of a deceased worker who died for lack of appropriate safety equipment to put you out of business. In other instances, you could be faced with fines and/or jail time for failing to follow OSHA and/or EPA industry regulations. In a "best case" scenario, the construction company you are providing services to may notice that you are not meeting appropriate safety standards and boot you from the construction site. You then stand to lose not only the contract you had for the job but also the possibility of getting future jobs with the company in question.

So make safety a high priority in your company. Make sure that your technicians know how to handle all potentially dangerous situations appropriately. If there is a question regarding safety procedures, make sure that someone in your company is available to answer the question accurately. Providing training and the right equipment does not cost nearly as much as dealing with a serious on-the-job accident brought about by ignorance or the desire to reduce overhead costs by a few measly percentage points.

A focus on safety shows that you care about your workers. While inexperienced HVAC technicians may be willing to put up with a company that fudges safety guidelines, experienced workers are aware of safety requirements and won't want to work for a company that does not care about their wellbeing. By following or even exceeding safety requirements, you show your employees that you have integrity and are keeping their best interests at heart.

Chapter Summary

As you can see, compensation can come in in a variety of offerings, and are important to recruiting and retention. Establishing a compensation program that is competitive and relevant in today's climate is important. It is equally important to include elements that provide future earnings for your employees like performance bonuses, profit sharing, or 401k programs. These programs give employees skin in the game. They're directly tied to how well they perform and protect the company's best interests.

Insurance benefits are becoming more common place with HVAC contractors. Health, life and dental insurance relieve employees from the stressors unexpected illnesses or accidents can bring by knowing that their out of pocket costs are mitigated. Secure and happy employees make great and loyal employees. Compensation is more than just wages. Compensation packages that recruit the best talent always include competitive wages and comprehensive benefit packages. We recommend your compensation programs be on par or better than your competition.

Scott Ritchey with Gary Kerns

Chapter 11
Knowledge is Confidence: In Their Own Words

Chapter Overview

In this chapter, you will learn:
- How to trust the data
- Thoughts on decision making
- Perspectives from fellow contractors

We've shared a lot of information in this book. If you've read the entire book in one sitting or even a few sittings, you may feel positively overwhelmed by all the information we shared. If you do, don't worry; it's a common reaction to reading through a lot of very specific data.

The reason we are sharing so much actionable information is not to overwhelm you; it's because we have seen through decades of consulting experience that knowledge is power. If you know what to do, when to do it, and how to do it, you'll be ready to take on any challenges you may face in the months and years ahead.

If you don't know something you need to know to run your business, you can still be successful if you are willing to get help from experienced professionals such as accountants, lawyers, business managers, or even your suppliers. You don't have to know everything there is to know to be successful if you allow others to help you handle aspects of your business that you have a hard time handling yourself. The problem arises when you don't have the necessary business skills needed to stay on top of your business and you fail to learn these skills or get help from others because you don't want to expose a business acumen deficiency or are too busy to even notice that you're not keeping track of your business income and expenses.

Working as a HVAC service technician is a great starting place for any HVAC business owner. You won't know what potential clients want or how to help them if you don't have training and/or experience as an HVAC technician. Even so, years or even decades of experience working in the HVAC industry will not equip you with all the skills and information you need to run a successful company. Your company must have a solid financial foundation to be successful, both now and in the future. Furthermore, this foundation must be built on financial pillars of success such as knowing the difference between margin and markup, understanding how overhead costs affect your net profits, knowing how to calculate job costs so you earn money on your work instead of losing it, and knowing how to read and understand the information presented in a profit and loss statement. You also need to know how to apply for and manage business credit, how to departmentalize the different segments of your HVAC company, and how to use the information on your income statement to accurately analyze your company.

We also believe the information we have presented on other topics such as market caps, service technician compensation, and overhead reduction are important to your success. However, we understand that applying this information is not a one-time job. You need to stay abreast of industry and local market trends to know what services to offer, what type of technicians to hire, and how to target your advertising so it reaches the right audience.

If you've read this book and realize that your HVAC company is slowly but surely heading downhill because of financial mismanagement, lack of direction, or any other reason, don't despair! After all, you're in good company. However, don't simply resign yourself to the situation as it is, accepting defeat and resolving that running an HVAC company is not for you. You can turn your HVAC company around or, in a worst-case scenario, start a new, successful one. The three steps outlined below can help you achieve success in this field even if your company is in dire straits.

Believe the Data

It is imperative that you believe the data presented in this book. If you don't, I hope that presenting this information will at least give you an interest in finding the truth for yourself. If you think that I am exaggerating the rate of failure for new HVAC startups, the difficulty of finding good HVAC workers, or any other statistics that I have presented in this book, do your own research. I suggest you look at established, authoritative industry resources such as the Bureau of Labor Statistics website, Achrnews.com (a website dedicated to

HVAC industry news), and the other sites that I have provided links to in this book.

If you analyze the data with an open mind, you will see that everything I am telling you here is the truth. The only question that remains is what you will do with it. You can ignore the truth, but that won't make it go away. Alternatively, you can use the facts and data here to blaze a pathway to eventual success.

It is even more important that you believe the data you see with your own eyes on your income statement and profit and loss statement. If something doesn't add up or doesn't look right, investigate further. It may be that there is a mistake on your statement and your company's financial standing isn't as bad as you think it is. However, in my experience, that's not usually the case. If your data is showing that your business is losing money even though you are getting multiple jobs and your workers are busy all the time, that is a warning signal. Take the data seriously because it will catch up to you if you don't.

The Confidence to Make Decisions and Implement Them

There are plenty of decisions you will need to make in your business even if things are going well. How do you get the confidence to make the right choices? I believe an analogy by well-known author and speaker Randy Nelson, "The Second Decision: The Qualified Entrepreneur," will help you get a good handle on the seemingly complex process of effective decision making.

Randy explains that, as a former Navy officer, he often spent months at a time in a submarine. A submarine travels underwater; there are no windows and no line of sight to the outside world. Even if you could see out a window in a submarine, all you would see is water, seaweed, and marine life. How do you know if you are heading in the right direction? As Randy accurately points out, you rely solely on metrics, gauges, and systems. You know if you keep your systems running well, they will let you know where you are going and how long it will take before you get there. The systems will also let you know if there is danger ahead.

At the same time, crew members also must rely on their own expertise and training. Decisions often have to be made in less than ideal circumstances without as much information as the captain would like. In such instances, those in charge assess what they do know, use the training they have to come to a decision, and then execute the decision. It's as simple and as effective as that.

It's not hard to see how this applies to the HVAC business world. As HVAC business owners, we simply cannot know everything we would like to about our business. I truly believe that knowing how to manage your finances and workers will lead to your long-term success, but I also readily acknowledge that changing markets, variables in supply and demand, increased competition, a slowdown in the housing market, and a plethora of other factors you don't see coming can throw your company off course. Technological changes in the industry can also affect how you do business. In short, there are a lot of unknowns in the HVAC business world.

However, you can't let the unknowns stop you from making decisions. In fact, failing to decide is a decision in itself. It is a choice to let go of the controls and allow your company to cruise on as it has been while you keep your fingers crossed in the hope that everything will turn out all right in the end. Allow me to tell you from personal experience that running a business in this way does not guarantee success.

Given that you can't predict the future, how can you make good decisions? The answer is quite simple. Just as a submarine crew must rely on gauges, systems, and data, you must rely on your financial statistics and industry benchmarks to know if your company is heading in the right direction. If the numbers aren't adding up as they should be, then you know something is wrong. If the numbers do add up, you know that your company will be able to withstand whatever the market may throw at it.

This is proven by five contractors who have agreed to tell their stories of how knowledge led to a confidence that helped them achieve their ultimate business goal of owning and operating a successful HVAC contracting business.

In Their Own Words

Herb Hovey, Owner, HH Hovey

My name is Herb Hovey, and I started in the HVAC industry in 1979. In 1990, after managing a local company as the vice president, I struck out on my own with literally one truck and one man. Like most contractors who start an HVAC company, I had all the technical experience one would need to operate my heating and cooling business.

I first met Scott Ritchey through the Climatic Corporation, the HVAC equipment distributorship that I bought equipment from and which employed Scott. At the time, Scott was putting together different programs that intrigued me. These programs were business classes for HVAC contractors and I subsequently applied to every single one.

Being in a big town, I had a lot of competition. A lot of that competition came from large, well-established companies and I needed to be able to compete. I was fine on the service repair and install side but needed guidance on the financial side of my business. After using what I learned in Scott's classes and programs, my business more than doubled year after year over the next several years to where it is today. Now, 25 years later, my company is one of the big guys in the market. One of the reasons for that was the value I was shown of having a robust service agreement business.

Once I was shown the per-dollar value of a service agreement, I was sold. Once the business hit the 500 quantity threshold, I was amazed at how fast my revenues grew. Scott would say that a service agreement is worth about $650 to $750 when you meet that threshold. Believe me, I had my doubts until he showed the math. Roughly 7% of my agreements need an air conditioner and 5% need a furnace every year. For my company, that is roughly 300 installations and an additional 500 demand service calls from contract customers who have breakdowns—all from my service agreement base of customers.

But the most important thing to me about my maintenance contract program is the strong customer relationships I have with my customers—having them helps carry us through the slow times of the year.

Thanks to Scott's information and guidance, my small one-truck business has grown into a five-truck service department and a three-truck installation department that employs 18 people. As I prepare to transition my business to my son, Michael, I have passed on Scott's teachings to him so he can have generational success. I've worked with Scott for more than 25 years, and his books and strategies got me to where I am today. I came to him to learn how to compete with the big boys, and now I am one of them.

Scott Ritchey with Gary Kerns

Rich Mullins, Owner, H2O Plumbing

My name is Rich, and I'm a plumbing entrepreneur. I started H2O Plumbing in 2003 as a one-man show, not looking to expand, and ran my business as such for 12 years. As family circumstances changed, my desire and need to grow changed as well. One day, I was doing some plumbing work for a high school friend of mine who was a heating and air conditioning contractor. Having known Brian for a long time, I watched his company grow from a small company startup like mine to being one of the larger contractors in the area. I asked Brian how he did it—how did his business become so large? I wanted to grow mine as well. Brian said, "I know someone who's going to change your life." That's when I placed a call to Scott Ritchey, and that call did indeed change my business life.

My first meeting with Scott was very eye-opening. He started by asking me where I was and where I wanted to be in the future. In the discussion, I was asking questions like, "Should I spend more money advertising or hire more people?" and out of the blue, Scott would ask, "How do you price your work and what is your overhead?" Being that I *was* the company, I told Scott that I did not have much overhead and gave him a number. He then said, "I guarantee you have more overhead than that." He asked if my shop was at my house and I said yes. Then he asked if I had a mortgage. I laughed and said yes, and he said, "You have more overhead than you think."

From there he showed me where I had these additional, hidden overhead costs and told me that if I wanted to have a larger company, I needed to act and price like a larger company. After discussing overhead some more, and showing me what the industry averages were, he asked me to price a job with him.

That's when he showed me that my pricing was inadequate and introduced the single-factor method of pricing jobs. Before our first meeting ended, Scott asked whether I was working on a job right now so we could try his pricing method. I said, "As a matter of fact, I am going to give a woman a bid tonight." He said, "You price it your way and I will price it my way—then let's compare."

When we compared the difference, Scott was, like, $1,100 more than I was. He said, "Use my price." I admit I was skeptical, but what did I have to lose? I gave the woman that price and she didn't even bat an eye—she just said "do it." Scott called me the next day to ask how it went and I told him I got the job. He said to have confidence in his method, as it works. Because of Scott's single-factor pricing method, I was able to give higher bids and was surprised that people were willing to pay more.

Scott and I had a few other meetings during that year discussing his throughput and benchmarking ideas. By applying Scott's business principles, I tripled my business that year, ending the year with $300,000 in sales and an additional employee. I have been working with Scott for two years now, and my business revenues for 2017 will be eight times greater, at $800,000 with three trucks. The plan was to work toward a million in three years and I have no doubt that will happen in this coming year.

Two fundamental things I learned from Scott were the importance of pricing and appearance. He pointed out that plumber or not, a businessman should dress for success and present every aspect of his business in the same way. Scott is always easy to reach and willing to talk. A lot of us guys are tradesmen to start; we don't know the business side. With Scott's help, all that can come into sync. If I had a friend who was struggling with his business like I was, I would say the same thing that was said to me: "I know a guy who will change your life."

Brian Schneider, Owner, Allegiance Heating & Air

I'm Brian Schneider, and I am an HVAC entrepreneur. Coming from a blue-collar family, I was never offered any business knowledge or experience. Growing up, I knew I wanted something different. I wanted something bigger than just clocking in for someone else. I wanted to change my family tree, have something bigger and better to leave behind for my children. My brother was in heating and air and got me interested in the industry. Being a calculated risk taker, it did not take long for me to decide to open my own business and begin the journey of changing my family tree.

I was doing business with a local distributor and had revenues around $600,000 when a friend of mine who worked for Plumbers Supply as a territory manager showed me how he and Plumbers Supply could help my business grow further and help me reach my goals of providing more for my family and employees.

Like most contractors, I was good on the technical side of the heating and air service but needed more on the business side of business. I decided to take Scott's Business 101 class, as the class was big on marketing, brand awareness, pricing, and throughputs. With the aid of these strategies, I took my business from $600,000 to its current multimillion-dollar volume, and I'm well on my way to my goal of changing not only my family tree but the family tree of my employees as well, as I am able to offer them more than most companies because of my higher profits.

Scott Ritchey gave me the business know-how I needed to grow my business. To me, this was not only about my family tree, but also about realizing that the business is bigger than just the business; it's about all the people that the business will touch and provide for. I would advise people to take those steps toward business growth, take a Scott Ritchey class, and not to be afraid to open up about what you need to grow.

Werner Vankleef, Owner, Vankleef's Heating and Air

I'm Werner Vankleef and I am an HVAC entrepreneur. After graduating college, I started my career in a sales position at a heating and air company before starting my own business in 1998. Like everyone else, I started with one truck and a helper; we will be celebrating our 20th anniversary in 2018.

Before Scott Richey, my company was growing, but growing slowly. There are two sides to the HVAC thing: the heating and air part and the business part. For me, the business side was lacking. Scott's Business 101 class at Plumbers Supply introduced me to many business best practices and turned things around. After taking my own numbers to crunch in his class rather than using fictional numbers, I realized I wasn't charging enough. I learned about the single-factor method where markup doesn't equal margin and about market margin caps for the residential and commercial businesses. Knowing these tactics, I was able to raise my prices, and today enjoy high double-digit net profits. I also learned how solutions proposals help me to upsell in the residential replacement business.

After applying these strategies, my growth continued. In 2015, I did $1.2 million in sales. Last year, I did $1.5 million. This year, $1.7 million. That is a nice steady rate of growth, and we are currently preparing to expand into a 3,200-square-foot space.

Ultimately, the proper pricing methods I learned from attending the Business 101 class gave me the confidence to continue growing. I would say the biggest impact of Scott Richey's teachings, for me, was just better understanding the business side of business. If someone else found themselves in the same boat as me, I would tell them that they deserve to grow, that they deserve to make a good profit for good work, and they deserve a better lifestyle for themselves, as well as their employees. When you understand the business side of the HVAC contracting business you can have all of it.

Van Edwards, Owner, Edwards Refrigeration

I got into this business because I didn't want to go to college—I wanted to work with my hands. I got an associate's degree in HVAC technology and started my business in 1977. I started from scratch; I knew no one in HVAC. I bought a service truck, got some tools, and started working on fridges and small window units. Eventually, I grew into working on residential central units, then to commercial. I've been doing it for 41 years now.

Edwards Refrigeration is different from most HVAC companies. We do a little bit of everything: residential, commercial, light industrial, HVAC, electrical, plumbing, commercial refrigeration, gas piping, trenching, and stand-by generators.

At my first meeting with Scott, we talked about gross margins. At the time, I was at 13 or 15%. Scott was concerned that was too low, but I told him I already was the highest priced guy in town. Scott said, "Let's try pricing a job my way. If you lose the job, I'll pay you the gross margin you would have made."

As it turned out, I didn't even know what my overhead was at the time. I was smart in fixing things, but not in business. Until I met Scott, I was just charging whatever everyone else was, maybe a little bit more. Then Scott's Business 101 class opened my eyes to how little I knew. After that class, I was scared. I realized I had to change things. I had to operate my business differently.

I started tracking things and making systematic changes: departmentalizing the service from the installs, tracking what each service truck would do, what kind of money we needed to make, and how to price. I'll tell you, Scott can't fix an air conditioner, but he can sure look at your year-end financials and fix those. I started training staff and service guys so I could delegate, and then I had more time to spend on the business side of things. I created what we call a "cookbook" with our own flat rate for pricing.

I'll be passing this business down to my son, and now I really have something of value to pass on. We're still a small company, with 12 of us here. My business did $1.6 million last year and $1.8 million this year. We're showing growth every year. We're up to an average of $200,000-plus per service truck, which is better than the industry benchmark. We started as a 100% residential company, but today we have a very strong commercial business as well; that's where the big money is. That's what I learned, to go where the biggest profit was.

In 1996, I was asked to compete for the Medal of Excellence honor, part of a national competition among Bryant dealers. The competition evaluated each company's business practices and compared them to their peers. Scott helped me with my first business plan, and right out of the gate, we won the bronze medal. The next year, we got the gold. Once you've won the gold, you had to compete only every other year because they did not want back-to-back winners and the gold winners were the judges for the next year's competition. With the help of my office manager and what I learned from Scott's class on business plans, we won the gold a total of five times and were retired from the program with the distinguished Pinnacle award.

When I was establishing my business, my son remembers all the time I wasn't there, and he doesn't want that for himself. Now that my business is operating so smoothly, he won't have to make those sacrifices. Scott basically told me I had to choose to work with wrenches or learn to work on the business. I chose business, and that's what has brought me my financial success. I'm currently up to 20% net profit before taxes.

I would say to other business owners, if you don't know how to price or what your overhead is, you need someone like Scott Ritchey. He can look at your P&L and tell you right there where your problems are. Scott put me, and a lot of others like me, on the right track.

Chapter Summary

The greatest satisfaction that Gary and I have had throughout our careers has not been the success we have experienced or the money made. Our satisfaction truly comes from seeing other contractors become wildly successful because someone finally painted the big picture with them and provided detailed steps for them to execute the plan.

I want to give a special thanks to the contractors who allowed us to interview them and share their stories. It meant a great deal to Gary and me. These contractors took the information we provided and did the heavy lifting required to make our information and processes work in their companies. I want to stress this: these contractors were committed to succeeding in their businesses. Many did not have a background in the financial concepts we discuss, but they were determined to learn them, trust the information, and then apply it to the framework of their businesses—and in doing so, became quite successful. Yet for every contractor that we see who is extremely successful, we see dozens who struggle mightily.

Our thought is: do not struggle by choice. There are so many resources out there that can help your business thrive financially. The real challenge for many of you will be to look into the mirror and ask yourself, "Do I really have a firm grasp on these ideas and concepts? If I don't, what do I need to do about it?" We recommend that you work with experts like us or distributors that you partner with who offer these types of resources.

Scott Ritchey with Gary Kerns

Chapter 12
Time to Get Started and Just Do It!

Chapter Overview

In this chapter, you will learn:
- Recommendations to get started
- Setting goals and prioritizing your steps to action
- Don't fear failure—learn from it

Now that you have read about how other contractors—ones who perhaps were in the same shape you are today—have succeeded, it is time for you to take the knowledge provided, believe in the data, and execute with confidence. You *can* turn your company into an extremely profitable and fun business to own.

If you are now ready to take on the true responsibilities of an HVAC entrepreneur, then consider the following building blocks to get you started. We recommend joining peer groups, attending seminars, collaborating with other non-competing contractors, or simply working with us.

Self-Evaluate and Set Goals

You need to set quantifiable goals for your company. Put in simple terms, this means setting goals that you can measure to know if you are getting where you want to go. First, evaluate where your current business is coming from. Are you doing low-margin work for new home contractors or rental property owners, but want to pursue more lucrative segments of the HVAC business? Do you have the current talent level to pursue other segments like service, commercial, or residential replacement work? If you wanted to grow your residential replacement business, how would you do it? Do you have the cash to invest in a lead generation program?

Once you have completed your self-evaluation, you can now set specific goals around what you want to do. It could be a financial goal such as, "I want to earn X amount in gross or net profits from HVAC service work by this time next year." Alternatively, you could set a goal for how many replacement installation jobs you want to get per year, or set a revenue goal per service truck and start using the pendulum swing sales method.

In any case, an actionable goal is one that is easy to measure. You can sit down a year from now, look at your statistics, and see if you reached your goal or not. Even more importantly, you can keep track of your progress toward your goal throughout the year to see if you are on track to reach it and adjust accordingly.

I would advise you, however, to set no more than one or two goals so that you can focus completely on taking the actions necessary to achieve the result you are looking for. Remember to decide first on what type of business you want to have. Once you know what markets you want to focus on, you can set the necessary goals for success.

Prioritize

Naturally, you can't reach all your goals at once. That is why you need to prioritize the goals you want to reach. While all your goals are important, some will naturally be more important than others.

If your business is facing financial difficulties, I urge you to make proper financial management one of your first goals. For instance, you could make a high-priority goal of turning a certain amount of net profit on each job you take. If you have liabilities that threaten your business's future, make a high-priority goal of paying back the money you owe within a specified timeframe so that you are no longer strangled by high interest rates and the threat of losing your company.

If your company is doing well, you can afford to set more positive high-priority goals. Perhaps your high-priority goal would be to increase revenue in a certain department by a certain amount. Another worthy goal would be to successfully break into a new market and earn a certain amount in this market by a set time.

Once your reach your highest priority goal, move onto the second one, the third one, and so forth.

Gather as Much Information as Possible

You can't know everything, but you can gather as much information as possible. If you have been in business for at least a year or two, you will find you have plenty of information available on your financial statements. Use this information to see where you company is heading and what you need to do to turn it around—or help it do even better. Conduct benchmarking and performance tracking to gain a clear perspective of where the business is now and where it is going in the future.

You can even apply this point if you haven't started an HVAC company yet but are planning to do so. Important information to gather includes not only the ratios, benchmarks, and equations outlined in this book but also current industry conditions in your city. Find out how many home or commercial building companies are at work in your local area. Find out what the market cap for any given HVAC service is in your city. Find out how many HVAC companies operate in your city and what types of jobs they focus on. Remember, knowledge is power.

Just Do It

Don't put off making decisions. Once you have set goals, decide which goals to reach for first, gather all the information you can, and then just *go for it.*

Deciding will enable you to get started in the right direction. As you continue in this direction, you will find that your well-informed, timely decisions enable you to increase your chances of success. Alternatively, you may find that you need to tweak or adjust the decisions you have made to improve results or perhaps change course to pursue a more favorable outcome as you continue on your journey.

For instance, you may have decided that you will try to break into a certain market and offer new services; however, you have now found that the overhead costs for offering the services you wanted to provide is higher than you initially anticipated. At the same time, maybe you have discovered that not all the HVAC technicians in your company have the skill sets to offer the services you want to focus on, leaving you short of staff needing to add a few new technicians. You will then need to adjust your decision to reach your company goals.

Don't abandon a decision just because it doesn't seem to be turning out as you had hoped; at the same time, don't stick with a decision that you can see isn't helping your company grow as it should.

Accept That You Will Make Mistakes

Perhaps one of the biggest hindrances to executing decisions with confidence is fear. We fear making mistakes. We fear failure. We fear looking dumb in front of our employees, colleagues, and customers. We then let fear control us. Fear causes many HVAC contractors to double down on decisions when they are obviously the wrong ones. Fear also causes many contractors to avoid making decisions in the first place. Fear, in the end, will either prevent your business from growing or potentially shut it down altogether.

If you are worried about making the wrong decision, take confidence in the fact that there is no such thing as a 100% successful HVAC contractor. All HVAC business owners make wrong decisions and then must deal with the consequences. Perhaps you have hired someone and now see that the person in question is not a good fit for your company. Maybe you chose the wrong type of advertising, made bad investments in used vehicles, or chose overly expensive office space or insurance coverage for your company that is stretching your overhead without meeting your needs.

There are all sorts of wrong decisions you can make in this field, and many HVAC businesses have been ruined by bad decisions. However, you can't let fear of failure stop you from making decisions. Numerous HVAC companies have also run aground because of failure to make any decisions at all. Be aware that failure is always a risk, but the more informed you are, the less sting some of these decisions can have.

Of course, you can avoid failure in many circumstances by following the advice outlined in this book. If you adhere to good business practices, you will make wise decisions based on sound advice rather than using your gut instinct or what the HVAC business owner did at the last company you worked for.

I would also like to take a minute to remind you of the importance of stepping back from the company every so often to see the full picture. Take some time away from work to look at where your company is heading, how it is getting there, what your company's strengths are, and what your weaknesses are. If you are working on your business instead of letting the business run you, you usually catch bad decisions quickly, and doing so will make it easy for you to deal with the fallout and move on.

It's Time to Win!

As the old saying goes, "Winners do; losers dream." Mind you, it's great to have dreams; it's just not so great when those dreams never become reality because you never step out of your comfort zone to make them happen.

I don't know where you are right now. Perhaps you are just thinking about starting up a new company, but the advice I have given you in this book makes it look like a lot of hard work. I will admit that it takes a lot to run a successful company. You have some of the skills needed if you have been working in the HVAC industry for any length of time, but to be successful, you will need to learn new skills, usually business skills, as you go. If this situation describes you, I urge you not to let fear hold you back from an exciting new world that is just waiting to be discovered. The joys of owning your own business can far surpass the satisfaction you get working for someone else if your dream truly is to own a successful HVAC business. Owning a business is empowering and can provide you with the funds you need to not only get by but thrive financially.

Maybe you started an HVAC business some time back and it's not producing the results you thought you would have by now. As you read this book, you may now realize that you have not applied some of the advice that I'm sharing with you. Perhaps you confused margin with markup and so have been losing money or not making enough on your jobs. Maybe you failed to realize the importance of offering benefits to your workers and so most of your HVAC technicians don't have the training and certification needed to take on the profitable jobs that you want to bid for.

If this describes you, don't give up on your company just yet. There is hope that you can turn it around. If your technicians have a good work ethic, then there is no skill or certification they can't earn with your help. You can advertise for new work even without a lot of advertising money in your budget; if you don't believe me, just try doing an internet search for "free ways to promote your business online." The results will probably astound you. Furthermore, it doesn't cost anything to properly calculate your overhead and start pricing jobs at the market cap for your area.

In a worst-case scenario, you may need to close up shop and start over. It may seem awful, but remember that it's just a bend in the road, not the end of the road. You now have valuable experience under your belt and the practical training you need to make your next company a success. Look at your failed business ventures as a learning experience and move on.

On the other hand, maybe you are a successful HVAC business owner. You have made it past the four-year mark, you know how to manage your company's finances, and you have plenty of good technicians you can rely on to do a good job. That's great, and I sincerely congratulate you! Your company is one of the few HVAC companies that have proven that it is possible to run a successful HVAC business. At the same time, you can't afford to coast along on your current success. You need to grow, both personally as a business owner and as a business. Take some time to think about the advice in this book to see if you can apply any of it to your company to improve efficiency, raise net profits, or simply improve morale among your workers.

As I touched on earlier, the HVAC industry is changing at a rapid pace. There are new products, new services, and new ways of doing things that can boost your business in more ways than one. Never allow yourself to coast along on your success. Strive to continue learning more and improving your business to keep costs low, keep your employees happy, and continually offer the highest level of customer satisfaction possible.

In Closing

Both Gary and I have enjoyed so much success in our profession that it's our desire to give something back to our community. That is why we wrote this book and are offering to work together with territory managers and contractors like you to improve the industry one contractor at a time. We worked our way up in the HVAC industry just as you are and we have learned the ins and outs of running a successful company through education and personal experiences that have highlighted the need for all companies to have a successful financial base.

I'll never forget the time many years ago when an HVAC business owner named Johnny approached me and asked me to buy his HVAC company for $50,000. I had a look at his company's financial statements and immediately realized that the company had a lot of potential but was losing money due to the simple fact that Johnny was not departmentalizing different aspects of his business and therefore was losing money on certain types of jobs. Buying the company would be a huge financial windfall for me, as I knew that I could turn it around in short order and probably sell it for a million dollars a few years down the line.

However, one thing stopped me from doing that. I talked with my mentor about the deal and he told me something I will never forget. He said, "Scott, if you take this deal, I never want to talk to you again. I never would have thought you would be one to take advantage of another human being."

My mentor's comments completely changed the way I looked at the deal. Instead of buying the business, I sat Johnny down and told him I wanted to turn his company around. Johnny insisted that he just wanted to sell. He had inherited the company from his dad and didn't really have an interest in figuring out how to manage a successful HVAC company. He just wanted to sell it and find some other way to earn a living. Even so, I showed him the simple financial mistakes he was making in calculating the overhead costs for his business; he was surprised to realize that his inaccurate cost calculations were causing him to price jobs at a loss. In the case of his business, the solution wasn't complicated or hard to implement; he simply needed to separate the installation work from his service work, accurately figure out the overhead for each department, and then calculate the price tag he put on each of his jobs using the accurate overhead figures.

A year later, I was driving to Florida for Thanksgiving when Johnny gave me a call. From the excitement in his voice, I knew why he was calling—plus, I knew his fiscal year ended in September and he was meeting with his accountant. Johnny's first words were, "Do you think the IRS is listening?" I laughed and asked, "How good was it?"

In one year's time, Johnny went from losing $400 for the year to making $127,000. Several years after that, I heard from Johnny again and found out that he had been able to fulfill his dream of purchasing a 200-acre horse farm. He is now thoroughly enjoying the fruits of his success along with his two daughters. That, to me, is worth far more than the $1 million in profit that I could have made from buying his business and flipping it a few years down the line.

It's has always saddened me and Gary to see the high failure rate for HVAC companies. We know how hard many contractors work to first learn the trade and then to start their own business. It's gut-wrenching to see contractors fail and not even know the cause of their failure. Many think they need to move to a large city where there are plenty of clients; others think they need to branch out into a new market, while yet others are tempted to blame market circumstances such as low demand for new housing for their failure to thrive.

If you think that outside circumstances and conditions are solely responsible for your inability to generate the money you need to stay in business, I would like you to take a close look at Gary's successful HVAC company. Gary has never laid a single worker off in more than 25 years. His company is in the small town of Richmond, Kentucky; it does not have hundreds of thousands of potential clients like a large metropolitan city. Though he may take jobs in nearby large cities, nearly all his work is generated right in his local area. And talk about making decisions to change your business! As I stated earlier in the book, he recently let more than $600,000 in annual revenues go because he wanted to focus more on residential installation, maintenance, and service instead of continuing his involvement with commercial construction jobs. That decision paid off greatly when he used our service technician sales strategies and generated over a million dollars with three service trucks.

Gary's company has thrived right where it is because he is willing to continually make changes as often as they are needed to stay in step with the market. Gary no longer works in the field; in fact, he probably hasn't touched a set of tools for more than 10 years. However, he is continually busy keeping up with financial records, market trends, and business statistics. He knows where he wants his company to be and his business information is showing him how to get there.

Gary also makes it a point to treat his employees with the respect and dignity they deserve. He has always guaranteed his workers a 40-hour work week because he knows that HVAC technicians, like the rest of us, cannot survive long with unstable work hours. While some technicians do thrive as independent contractors, Gary and I have found from our extensive experience that you cannot retain full-time workers if you can't offer them a full-time job position with all the benefits that such a position should bring. At present, Gary's company offers his employees full benefits. These include not only health insurance but also dental and vision insurance, phones, uniforms and even 401(k) plans.

As it turns out, we are not smarter than you. We're not more capable or more experienced. The HVAC business owners who have turned their companies around are also just like you. The only difference is that we have learned simple principles of success that every entrepreneur needs to know to run any type of successful company. Now we are teaching these principles to you so you can enjoy success and all the many benefits that come with it.

Our hope is that, in a few years' time, our industry will look very different than it does now as HVAC contractors all over the country learn how to create successful companies that provide great services to employees and customers alike.

About the Authors

Scott Ritchey

Author Scott Ritchey is on a mission to help contractors across the country. He focuses his career on providing sound financial training that will help any HVAC owner sustain a successful business. To date, Scott has worked with 395+ companies and thousands of individuals, enabling them to learn important financial principles that all but guarantee success in the HVAC contracting industry.

Scott has nearly three decades of experience consulting with HVAC companies all over the United States, and the success rate of the companies he's worked with is simply astounding.

Gary Kerns

Contributing Author Gary Kerns is president and owner of Superior Heating and Air Conditioning, Inc. His background has given Gary a unique outlook on the HVAC contracting industry. After high school, he joined the Air Force.

After his stint in the Air Force, Gary became an apprentice with a commercial HVAC company. It was a perfect fit. Gary's talent as a top-notch service mechanic came to the fore and he became enamored with the idea of starting his own HVAC company and did.

CPSIA information can be obtained
at www.ICGtesting.com
Printed in the USA
LVHW080339240920
666973LV00006B/330